Photonic Crystals

Photonic Crystals

Molding the Flow of Light

John. D. Joannopoulos

Robert D. Meade

Joshua N. Winn

PRINCETON UNIVERSITY PRESS

Library of Congress cataloguing-in-Publication Data

Joannopoulos, J. D. (John D.), 1947–
Photonic crystals: molding the flow of light / by John D. Joannopoulos,
Robert D. Meade, Joshua N. Winn.
p.   cm.
Includes bibliographical references and index.
ISBN 0-691-03744-2
1. Photons.   2. Crystal optics.   I. Meade, Robert D. (Robert David), 1963–
II. Winn, Joshua N.   III. Title.
QC793.5.P427J63   1995
548′.9—dc20   94-12372

To Kyriaki and Lynne

"If only it were possible to make dielectric materials in which electromagnetic waves cannot propagate at certain frequencies, all kinds of almost-magical things would be possible."

—John Maddox

Nature **348,** 481 (1990)

Contents

Acknowledgments

It is always difficult to write a book about a topic that is still a subject of active research. Part of the challenge lies in translating research papers directly into a text. Without the benefit of decades of classroom instruction, there is no existing body of pedagogical arguments and exercises to draw from.

Even more challenging is the task of deciding which material to include. Who knows which approaches will withstand the test of time? It is impossible to know, so in this text we have tried to include only those subjects of the field which we consider most likely to be timeless. That is, we present the fundamentals and the proven results, hoping that afterwards the reader will be prepared to read and understand the current literature. Certainly there is much to add to this material as research continues, but we have tried to take care that nothing need be subtracted. Of course this has come at the expense of leaving out new and exciting results which are a bit more speculative.

If we have succeeded in these tasks, it is only because of the assistance of dozens of colleagues and friends. In particular, we have benefited from collaborations with Oscar Alerhand, G. Arjavalingam, Karl Brommer, Shanhui Fan, Ilya Kurland, Andrew Rappe, Bill Robertson, and Eli Yablonovitch. We also thank Paul Gourley and Pierre Villeneuve for their contributions to this book. In addition, we gratefully thank Tomas Arias and Kyeongjae Cho for helpful insights and productive conversations. Finally, we would like to acknowledge the partial support of the Office of Naval Research and the Army Research Office while this manuscript was being prepared.

Photonic Crystals

1

Introduction

Controlling the Properties of Materials

Many of the true breakthroughs in our technology have resulted from a deeper understanding of the properties of materials. The rise of our ancestors from the Stone Age through the Iron Age is largely a story of humanity's increasing recognition of the utility of natural materials. Prehistoric people fashioned tools based on their knowledge of the durability of stone and the hardness of iron. In each case, humankind learned to extract a material from the Earth whose fixed properties proved useful.

Eventually, early engineers learned to do more than just take what the Earth provides in raw form. By tinkering with existing materials, they produced substances with even more desirable properties—from the luster of early bronze alloys to the reliability of modern steel and concrete. Today we boast a collection of wholly artificial materials with a tremendous range of *mechanical* properties, thanks to advances in metallurgy, ceramics, and plastics.

In this century, our control over materials has spread to include their *electrical* properties. Advances in semiconductor physics have allowed us to tailor the conducting properties of certain materials, thereby initiating the transistor revolution in electronics. With new alloys and ceramics, scientists have invented high-temperature superconductors. It is impossible to overstate the impact that our advances in these fields have had on our society.

In the last decade a new frontier has emerged with a similar goal: to control the *optical* properties of materials. If we could engineer materials that prohibit the propagation of light, or allow it only in certain directions at certain frequencies, or localize light in specified areas, our technology would benefit. Already, fiber-optic cables, which simply guide light, have revolutionized the telecommunications industry.

Lasers, high-speed computers, and spectroscopy are just a few of the fields next in line to reap the benefits from the advances in optical materials. It is with these goals in mind that this book is written.

Photonic Crystals

What sort of material can afford us complete control over light propagation? To answer this question, we rely on an analogy with our successful electronic materials. A crystal is a periodic arrangement of atoms or molecules; that is, a crystal lattice results when a small, basic building block of atoms or molecules is repeated in space. A crystal therefore presents a periodic potential to an electron propagating through it, and the geometry of the crystal dictates many of the conduction properties of the crystal.

In particular, the lattice might introduce gaps into the energy band structure of the crystal, so that (due to Bragg-like diffraction from the atoms) electrons are forbidden to propagate with certain energies in certain directions. If the lattice potential is strong enough, the gap might extend to all possible directions, resulting in a *complete band gap*. For example, a semiconductor has a complete band gap between the valence and conduction energy bands.

The optical analogy is the *photonic* crystal, in which the periodic "potential" is due to a lattice of macroscopic dielectric media instead of atoms. If the dielectric constants of the materials in the crystal are different enough, and the absorption of light by the material is minimal, then scattering at the interfaces can produce many of the same phenomena for *photons* (light modes) as the atomic potential does for *electrons*. One solution to the problem of optical control and manipulation is thus a *photonic crystal,* a low-loss periodic dielectric medium. In particular, we can design and construct photonic crystals with *photonic band gaps,* preventing light from propagating in certain directions with specified energies.

To develop this concept, consider how two different devices already in common use—metallic waveguides and dielectric mirrors—relate to photonic crystals. Metallic cavities and waveguides are widely used to control microwave propagation. A metallic cavity does not allow electromagnetic waves to propagate below a certain threshold frequency, and a metallic waveguide only allows propagation along its axis. Both of these properties are useful, and would also be extremely useful at frequencies outside the microwave regime.

However, electromagnetic waves at other frequencies (visible light, for instance) are quickly dissipated in metallic components, which makes this method of optical control impossible to generalize. Photonic crystals can not only mimic the properties of cavities and waveguides, but are also scalable and applicable to a wider range of frequencies. We may construct a photonic crystal of a given geometry with millimeter dimensions for microwave control, or with micron dimensions for infrared control.

Another widely used optical device is the dielectric mirror, the familiar "quarter-wave stack" of alternating layers of different dielectric materials. Light of the proper wavelength, when incident on such a layered material, is completely reflected. The reason is that the light wave is scattered at the layer interfaces, and if the spacing is just right, the multiply-scattered waves interfere destructively inside the material. This effect is well known—it forms the basis of many devices, including dielectric mirrors, dielectric Fabry-Perot filters, and distributed feedback lasers. All contain low-loss dielectrics that are periodic in one dimension, so by our definition they are one-dimensional photonic crystals. However, while such mirrors are tremendously useful, they only reflect light at normal or near-normal incidence to the layered material.

If, for some frequency range, a photonic crystal reflects light of *any* polarization incident at *any* angle, we say that the crystal has a *complete photonic band gap*. In such a crystal, no light modes can propagate if they have a frequency within that range. A simple dielectric mirror cannot have a *complete* band gap, because scattering occurs only along one axis. In order to create a material with a complete photonic band gap, we must arrange the contrasting dielectrics in a lattice that is periodic along three axes.

An Overview of the Text

The aim of this text is to provide a comprehensive description of the propagation of light in photonic crystals. To accomplish this, we discuss the properties of photonic crystals of gradually increasing complexity—beginning with one-dimensional crystals and moving on to the more intricate and useful properties of two- and three-dimensional systems (see fig. 1). After arming ourselves with the appropriate theoretical tools, we begin to address pragmatic questions: which structures yield what properties, and why?

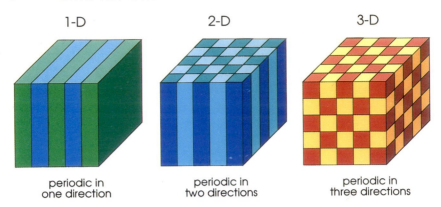

1-D   2-D   3-D

periodic in   periodic in   periodic in
one direction   two directions   three directions

Figure 1 Simple examples of one-, two-, and three-dimensional photonic crystals. The different colors represent materials with different dielectric constants. The defining feature of a photonic crystal is the periodicity of dielectric material along one or more axes.

This text is designed for a broad audience. The only strict prerequisites are familiarity with the macroscopic Maxwell equations and the notion of harmonic modes. From these two building blocks, we develop all of the mathematical and physical tools that we need. Interested undergraduates will find the text approachable, yet professional researchers might also use our heuristics and results to choose photonic crystals for their own applications. Readers who are familiar with quantum mechanics and solid-state physics are at some advantage, because our formalism owes a great deal to the techniques and nomenclature of those fields—appendix A explores this analogy in detail.

The field of photonic crystals is a marriage of solid-state physics and electromagnetism. Crystal structures are citizens of solid-state physics, but in photonic crystals the electrons are replaced by electromagnetic waves. Accordingly, we present the basic concepts of both subjects before launching into an analysis of photonic crystals. In chapter 2, we discuss the macroscopic Maxwell equations as they apply to dielectric media. These equations will be cast as a single Hermitian differential equation, a form in which many useful properties become transparent: the orthogonality of modes, the electromagnetic variational theorem, and the scaling laws of dielectric systems.

Chapter 3 presents some basic concepts of solid-state physics and symmetry theory as they apply to photonic crystals. It is common to apply symmetry arguments to understand the propagation of electrons in a periodic crystal potential. Similar arguments also apply to the case of light propagating in a photonic crystal. We examine the consequences of translational, rotational, mirror-reflection, inversion, and time-reversal symmetries in photonic crystals, while introducing some terminology from solid-state physics.

To develop the basic notions underlying photonic crystals, we begin by reviewing the properties of one-dimensional photonic crystals. In chapter 4, we will see that one-dimensional systems can exhibit three important phenomena: photonic band gaps, localized modes, and surface states. However, because the index contrast is only along one direction, the band gaps and the bound states are limited to that direction. Nevertheless, this simple and traditional system illustrates most of the physical features of the more complex two- and three-dimensional photonic crystals.

In chapter 5, we discuss the properties of two-dimensional photonic crystals, which are periodic in two directions and homogeneous in the third. These systems can have a photonic band gap in the plane of periodicity. By analyzing field patterns of some electromagnetic modes in different crystals, we gain insight into the nature of band gaps in complex periodic media. We will see that defects in such two-dimensional crystals can localize modes in the plane, and that the faces of the crystal can support surface states.

Chapter 6 addresses three-dimensional photonic crystals, which are periodic along three axes. It is a remarkable fact that such a system can have a *complete* photonic band gap, so that no propagating states are allowed in any direction in the crystal. The discovery of particular dielectric structures which possess a complete photonic band gap was one of the most important achievements in this field. These crystals are sufficiently complex to allow localization of light at point defects and propagation along linear defects.

Finally, in the last chapter, we use the tools and ideas that were introduced in previous chapters to design some simple optical components. Specifically, we will sketch designs for a reflecting dielectric, a resonant cavity, and a dielectric waveguide. The "walk-through" examples are intended not only to illustrate the device applications of photonic crystals, but also to provide a brief review of the material contained elsewhere in the text.

2

Electromagnetism in Mixed Dielectric Media

In order to study the propagation of light in a photonic crystal, we must turn to the Maxwell equations. After specializing to the case of a mixed dielectric medium, we cast the Maxwell equations as a linear Hermitian eigenvalue problem. From this formulation, in close analogy with the Schrödinger equation of quantum mechanics, comes a variety of useful properties, including the orthogonality of modes and the electromagnetic variational theorem. Finally, we show how electromagnetic problems with different overall length and dielectric scales can be related.

The Macroscopic Maxwell Equations

All of macroscopic electromagnetism, including the propagation of light in a photonic crystal, is governed by the four macroscopic Maxwell equations. In cgs units, they are

$$\nabla \cdot \mathbf{B} = 0 \qquad \nabla \times \mathbf{E} + \frac{1}{c}\frac{\partial \mathbf{B}}{\partial t} = 0$$

$$\nabla \cdot \mathbf{D} = 4\pi\rho \qquad \nabla \times \mathbf{H} - \frac{1}{c}\frac{\partial \mathbf{D}}{\partial t} = \frac{4\pi}{c}\mathbf{J}, \tag{1}$$

where (respectively) $\mathbf{E}$ and $\mathbf{H}$ are the macroscopic electric and magnetic fields, $\mathbf{D}$ and $\mathbf{B}$ are the displacement and magnetic induction fields, and ρ and $\mathbf{J}$ are the free charges and currents. An excellent derivation of these equations from their microscopic counterparts is given in Jackson (1962).

We will restrict ourselves to propagation within a *mixed dielectric medium,* a composite of regions of homogeneous dielectric material, with no free charges or currents. This composite need not be periodic,

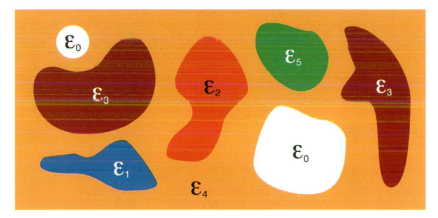

Figure 1 A composite of macroscopic regions of homogeneous dielectric media. There are no charges or currents. Although $\varepsilon(\mathbf{r})$ in equation (1) is arbitrary, most of our attention will focus on materials with patches of homogenous dielectric, like the one above.

as illustrated in figure 1. With this type of medium in mind, in which light propagates but there are no sources of light, we can set $\rho = \mathbf{J} = 0$.

Next we relate $\mathbf{D}$ to $\mathbf{E}$ and $\mathbf{B}$ to $\mathbf{H}$ with the constitutive relations appropriate to our problem. Quite generally, the components D_i of the displacement field $\mathbf{D}$ are related to the components E_i of the electric field $\mathbf{E}$ via a complicated power series, as in Bloembergen (1965):

$$D_i = \sum_j \varepsilon_{ij} E_j + \sum_{j,} k \, \chi_{ijk} E_j E_k + O(E^3). \tag{2}$$

However, for many dielectric materials, it is a good approximation to employ the following standard assumptions. First, we assume the field strengths are small enough so that we are in the linear regime, so χ and all higher terms can be ignored. Second, we assume the material is macroscopic and isotropic, so that $\mathbf{E}(\mathbf{r}, \omega)$ and $\mathbf{D}(\mathbf{r}, \omega)$ are related by a scalar dielectric constant $\varepsilon(\mathbf{r}, \omega)$.[1] Third, we ignore any explicit frequency dependence of the dielectric constant. Instead, we simply choose the value of the dielectric constant appropriate to the frequency range of the physical system we are considering. Fourth, we focus only on low-loss dielectrics, which means we can treat $\varepsilon(\mathbf{r})$ as purely real.[2]

---

[1]It is straightforward to generalize this formalism to anisotropic media, in which $\mathbf{D}$ and $\mathbf{E}$ are related by a dielectric tensor ε_{ij}.

[2]Complex dielectric constants are used to account for absorption, as in Jackson (1962).

When all is said and done, we have $\mathbf{D}(\mathbf{r}) = \varepsilon(\mathbf{r})\mathbf{E}(\mathbf{r})$. An equation similar to (2) relates $\mathbf{B}$ to $\mathbf{H}$. However, for most dielectric materials of interest, the magnetic permeability is very close to unity and we may set $\mathbf{B} = \mathbf{H}$.

With all of these assumptions in place, the Maxwell equations (1) become

$$\nabla \cdot \mathbf{H}(\mathbf{r}, t) = 0 \qquad \nabla \times \mathbf{E}(\mathbf{r}, t) + \frac{1}{c}\frac{\partial \mathbf{H}(\mathbf{r}, t)}{\partial t} = 0$$

$$\nabla \cdot \varepsilon(\mathbf{r})\mathbf{E}(\mathbf{r}, t) = 0 \qquad \nabla \times \mathbf{H}(\mathbf{r}, t) - \frac{\varepsilon(\mathbf{r})}{c}\frac{\partial \mathbf{E}(\mathbf{r}, t)}{\partial t} = 0. \tag{3}$$

We have restricted ourselves to linear, lossless materials. The reader may wonder if we are missing out on interesting physical phenomena with these restrictions. This may be the case, but it is a remarkable fact that many interesting and useful properties arise from "simple" linear, lossless materials. In addition, the theory of these materials is much simpler to analyze at this level, and is practically exact. For these reasons, we will be concerned with this case for the rest of the text.

In general both $\mathbf{E}$ and $\mathbf{H}$ are complicated functions of time and space. But since the Maxwell equations are linear, we can separate out the time dependence by expanding the fields into a set of harmonic modes. In this and the following sections we will concern ourselves with the restrictions that the Maxwell equations impose on a field pattern that happens to vary sinusoidally (harmonically) with time. This is no great limitation, since we know by Fourier analysis that we can build *any* solution with an appropriate combination of these harmonic modes. Often we will refer to them simply as modes.

We employ the familiar trick of using a complex-valued field for mathematical convenience, remembering to take the real part to obtain the physical fields. This allows us to write a harmonic mode as a certain field pattern times a complex exponential:

$$\begin{aligned} \mathbf{H}(\mathbf{r}, t) &= \mathbf{H}(\mathbf{r})e^{i\omega t} \\ \mathbf{E}(\mathbf{r}, t) &= \mathbf{E}(\mathbf{r})e^{i\omega t}. \end{aligned} \tag{4}$$

To find the equations for the mode profiles of a given frequency, we insert the above equations into (3). The two divergence equations give the simple conditions:

$$\nabla \cdot \mathbf{H}(\mathbf{r}) = \nabla \cdot \mathbf{D}(\mathbf{r}) = 0. \tag{5}$$

These equations have a simple physical interpretation. There are no point sources or sinks of displacement and magnetic fields in the medium. Alternatively, the field configurations are built up of electromagnetic waves that are *transverse*. That is, if we have a plane wave $\mathbf{H}(\mathbf{r}) = \mathbf{a} \exp(i\mathbf{k} \cdot \mathbf{r})$, equation (5) requires that $\mathbf{a} \cdot \mathbf{k} = 0$. We can focus our attention on the other two of the Maxwell equations as long as we are always careful to enforce this transversality requirement.

The two curl equations relate $\mathbf{E}(\mathbf{r})$ to $\mathbf{H}(\mathbf{r})$:

$$\nabla \times \mathbf{E}(\mathbf{r}) + \frac{i\omega}{c} \mathbf{H}(\mathbf{r}) = 0$$

$$\nabla \times \mathbf{H}(\mathbf{r}) - \frac{i\omega}{c} \varepsilon(\mathbf{r})\mathbf{E}(\mathbf{r}) = 0. \tag{6}$$

We can decouple these equations in the following way. Divide the bottom equation of (6) by $\varepsilon(\mathbf{r})$, and then take the curl. Then use the first equation to eliminate $\mathbf{E}(\mathbf{r})$. The result is an equation entirely in $\mathbf{H}(\mathbf{r})$:

$$\boxed{\nabla \times \left(\frac{1}{\varepsilon(\mathbf{r})} \nabla \times \mathbf{H}(\mathbf{r}) \right) = \left(\frac{\omega}{c} \right)^2 \mathbf{H}(\mathbf{r})} \tag{7}$$

This is the master equation. In addition to the divergence equation (5), it completely determines $\mathbf{H}(\mathbf{r})$. Our strategy will be the following: for a given photonic crystal $\varepsilon(\mathbf{r})$, solve the master equation to find the modes $\mathbf{H}(\mathbf{r})$ for a given frequency, subject to the transversality requirement. Then use the second of (6) to recover $\mathbf{E}(\mathbf{r})$:

$$\mathbf{E}(\mathbf{r}) = \left(\frac{-ic}{\omega\varepsilon(\mathbf{r})} \right) \nabla \times \mathbf{H}(\mathbf{r}). \tag{8}$$

The reason why we chose to formulate the problem in terms of $\mathbf{H}(\mathbf{r})$ and not $\mathbf{E}(\mathbf{r})$ will be discussed in a later section of this chapter.

Electromagnetism as an Eigenvalue Problem

As discussed in the previous section, the heart of the Maxwell equations for a harmonic mode in a mixed dielectric medium is a complicated differential equation for $\mathbf{H}(\mathbf{r})$, given by equation (7). The content of the equation is this: perform a series of operations on a function $\mathbf{H}(\mathbf{r})$, and if $\mathbf{H}(\mathbf{r})$ is really an allowable electromagnetic mode, the result will just be a constant times the original function $\mathbf{H}(\mathbf{r})$. This situation arises often in mathematical physics, and is called an *eigenvalue problem*. If the result of an operation on a function is just the function itself, multiplied by some constant, then the function is called an *eigenfunction* or *eigenvector* of that operator, and the multiplicative constant is called the *eigenvalue*.

In this case, we identify the left side of the master equation as an operator Θ acting on $\mathbf{H}(\mathbf{r})$ to make it look explicitly like an eigenvalue problem:

$$\Theta\mathbf{H}(\mathbf{r}) = \left(\frac{\omega}{c}\right)^2 \mathbf{H}(\mathbf{r}). \tag{9}$$

We have identified Θ as the differential operator that takes the curl, then divides by $\varepsilon(\mathbf{r})$, and then takes the curl again:

$$\Theta\mathbf{H}(\mathbf{r}) \equiv \nabla \times \left(\frac{1}{\varepsilon(\mathbf{r})}\nabla \mathbf{H}(\mathbf{r})\right). \tag{10}$$

The eigenvectors $\mathbf{H}(\mathbf{r})$ are the field patterns of the harmonic modes, and the eigenvalues $(\omega/c)^2$ are proportional to the squared frequencies of those modes. One important thing to notice is that the operator Θ is a *linear* operator. That is, any linear combination of solutions is itself a solution; if $\mathbf{H}_1(\mathbf{r})$ and $\mathbf{H}_2(\mathbf{r})$ are both solutions of (9) with the same frequency ω, then so is $\alpha\mathbf{H}_1(\mathbf{r}) + \beta\mathbf{H}_2(\mathbf{r})$, where α and β are constants. For example, given a certain mode profile, we can construct another legitimate mode profile with the same frequency by simply doubling the field strength everywhere ($\alpha = 2$, $\beta = 0$). For this reason we consider two field patterns that differ only by an overall multiplier to be essentially the same mode.

Our operator notation is reminiscent of quantum mechanics, in which we obtain an eigenvalue equation by operating on the wave function with the Hamiltonian. A reader familiar with quantum mechanics might recall some key properties of the eigenfunctions of the Hamiltonian: they have real eigenvalues, they are orthogonal, they can be obtained by a variational principle, and they may be catalogued by their symmetry properties (see, for example, Shankar 1982).

All of these same useful properties hold for our formulation of electromagnetism. In both cases, the properties rely on the fact that the main operator is a special type of linear operator known as a *Hermitian* operator. In the coming sections we will develop these properties one by one. We conclude this section by showing what it means for an operator to be Hermitian. First, in analogy with the inner product of two wave functions, we define the inner product of two vector fields $\mathbf{F}(\mathbf{r})$ and $\mathbf{G}(\mathbf{r})$ as

$$(\mathbf{F}, \mathbf{G}) \equiv \int d\mathbf{r} \mathbf{F}^*(\mathbf{r}) \cdot \mathbf{G}(\mathbf{r}). \qquad (11)$$

Note that a simple consequence of this definition is that $(\mathbf{F}, \mathbf{G}) = (\mathbf{G}, \mathbf{F})^*$ for any $\mathbf{F}$ and $\mathbf{G}$. Also note that $(\mathbf{F}, \mathbf{F})$ is always real, even if $\mathbf{F}$ itself is complex. In fact, if $\mathbf{F}(\mathbf{r})$ is a harmonic mode of our electromagnetic system, we can always set $(\mathbf{F}, \mathbf{F}) = 1$ by using our freedom to scale any mode by an overall multiplier. Given $\mathbf{F}'(\mathbf{r})$ with $(\mathbf{F}', \mathbf{F}') \neq 1$, create

$$\mathbf{F}(\mathbf{r}) = \frac{\mathbf{F}'(\mathbf{r})}{\sqrt{(\mathbf{F}', \mathbf{F}')}}. \qquad (12)$$

From our previous discussion, $\mathbf{F}(\mathbf{r})$ is really the same mode as $\mathbf{F}'(\mathbf{r})$, since it differs only by an overall multiplier, but now (as the reader can easily verify) we have $(\mathbf{F}, \mathbf{F}) = 1$. We say that $\mathbf{F}(\mathbf{r})$ has been *normalized.* Normalized modes are very useful in formal arguments. If, however, one is interested in the physical energy of the field and not just its spatial profile, the overall multiplier is important.[3]

Next, we say that an operator Ξ is *Hermitian* if $(\mathbf{F}, \Xi\mathbf{G}) = (\Xi\mathbf{F}, \mathbf{G})$ for any vector fields $\mathbf{F}(\mathbf{r})$ and $\mathbf{G}(\mathbf{r})$. That is, it does not matter which function is operated upon before taking the inner product. Clearly not all operators are Hermitian. To show that Θ is Hermitian, we perform an integration by parts twice:

$$(\mathbf{F}, \Theta\mathbf{G}) = \int d\mathbf{r} \, \mathbf{F}^* \cdot \nabla \times \left(\frac{1}{\varepsilon} \nabla \times \mathbf{G} \right)$$

$$= \int d\mathbf{r} \, (\nabla \times \mathbf{F})^* \cdot \frac{1}{\varepsilon} \nabla \times \mathbf{G} \qquad (13)$$

$$= \int d\mathbf{r} \left(\nabla \times \left(\frac{1}{\varepsilon} \nabla \times \mathbf{F} \right) \right)^* \cdot \mathbf{G} = (\Theta\mathbf{F}, \mathbf{G}).$$

In performing the integrations by parts, we neglected the surface terms that involve the values of the fields at the boundaries of integration. This is because in all cases of interest, one of two things will be

---

[3]This distinction is discussed again after equation (23).

true: either the fields decay to zero at large distances, or the fields are periodic in the region of integration. In both cases, the surface terms vanish.

General Properties of the Harmonic Modes

Having established that Θ is Hermitian, we can now show that Θ must have real eigenvalues. Suppose $\mathbf{H}(\mathbf{r})$ is an eigenvector of Θ with eigenvalue $(\omega/c)^2$. Take the inner product of the master equation (7) with $\mathbf{H}(\mathbf{r})$:

$$\begin{aligned}
\Theta\mathbf{H}(\mathbf{r}) &= (\omega^2/c^2)\mathbf{H}(\mathbf{r}) \\
(\mathbf{H}, \Theta\mathbf{H}) &= (\omega^2/c^2)\,(\mathbf{H}, \mathbf{H}) \\
(\mathbf{H}, \Theta\mathbf{H})^* &= (\omega^2/c^2)^*\,(\mathbf{H}, \mathbf{H}).
\end{aligned} \tag{14}$$

Because Θ is Hermitian, we know that $(\mathbf{H}, \Theta\mathbf{H}) = (\Theta\mathbf{H}, \mathbf{H})$. Additionally, from the definition of the inner product we know that $(\mathbf{H}, \Xi\mathbf{H}) = (\Xi\mathbf{H}, \mathbf{H})^*$ for *any* operator Ξ. Using these two pieces of information, we continue:

$$\begin{gathered}
(\mathbf{H}, \Theta\mathbf{H})^* = (\omega^2/c^2)^*\,(\mathbf{H}, \mathbf{H}) = (\Theta\mathbf{H}, \mathbf{H}) = (\omega^2/c^2)\,(\mathbf{H}, \mathbf{H}) \\
(\omega^2/c^2)^* = (\omega^2/c^2)
\end{gathered} \tag{15}$$

It follows that $\omega^2 = \omega^{2*}$, or that ω^2 is real. By a different argument, we can also show that ω^2 is always positive. Set $\mathbf{F} = \mathbf{G} = \mathbf{H}$ in the middle equation of (13), to obtain:

$$(\mathbf{H}, \mathbf{H})\left(\frac{\omega}{c}\right)^2 = (\mathbf{H}, \Theta\mathbf{H}) = \int d\mathbf{r}\,\frac{1}{\varepsilon}|\nabla \times \mathbf{H}|^2. \tag{16}$$

Since $\varepsilon(\mathbf{r}) > 0$ everywhere, the integrand on the right-hand side is everywhere positive. Therefore all of the eigenvalues ω^2 must be positive, and ω is real.

Additionally, the Hermiticity of Θ forces any two harmonic modes $\mathbf{H}_1(\mathbf{r})$ and $\mathbf{H}_2(\mathbf{r})$ with different frequencies ω_1 and ω_2 to have an inner product of zero. Consider two normalized modes, $\mathbf{H}_1(\mathbf{r})$ and $\mathbf{H}_2(\mathbf{r})$, with frequencies ω_1 and ω_2:

$$\begin{gathered}
\omega_1^2(\mathbf{H}_2, \mathbf{H}_1) = c^2(\mathbf{H}_2, \Theta\mathbf{H}_1) = c^2(\Theta\mathbf{H}_2, \mathbf{H}_1) = \omega_2^2(\mathbf{H}_2, \mathbf{H}_1) \\
\rightarrow (\omega_1^2 - \omega_2^2)\,(\mathbf{H}_2, \mathbf{H}_1) = 0.
\end{gathered} \tag{17}$$

If $\omega_1 \neq \omega_2$, then we must have $(\mathbf{H}_1, \mathbf{H}_2) = 0$ and we say $\mathbf{H}_1$ and $\mathbf{H}_2$ are *orthogonal* modes. If two harmonic modes have equal frequencies $\omega_1 = \omega_2$, then we say they are *degenerate* and they are not necessarily orthogonal. For two modes to be degenerate requires what seems, on the face of it, to be an astonishing coincidence: two different field patterns that happen to have precisely the same frequency. Usually there is a symmetry that is responsible for the "coincidence." For example, if the dielectric configuration is invariant under a 120° rotation, modes that differ only by a 120° rotation are expected to have the same frequency. Such modes are degenerate and are not necessarily orthogonal.

However, since Θ is linear, any linear combination of these degenerate modes is itself a mode with that same frequency. As in quantum mechanics, we can always choose to work with linear combinations that *are* orthogonal (as in Merzbacher 1961, for example). This allows us to say quite generally that different modes are orthogonal.

A conceptual grasp of what it means to be orthogonal is best acquired in one dimension. What follows is a brief explanation (not mathematically rigorous, but perhaps useful to the intuition) that may help in understanding the significance of orthogonality. For two real one-dimensional functions $f(x)$ and $g(x)$ to be orthogonal means that

$$(f, g) = \int f(x)g(x)dx = 0. \tag{18}$$

In a sense, the product fg must be negative at least as much as it is positive over the interval of interest, so that the net integral vanishes. For example, the familiar set of functions $f_n(x) = \sin(n\pi x/L)$ are all orthogonal in the interval from $x = 0$ to $x = L$. Note that all have different numbers of nodes, where $f_n(x) = 0$. In particular, f_n has $n - 1$ nodes. The product of any two different f_n is positive as often as it is negative, so the inner product vanishes.

The extension to a higher number of dimensions is a bit unclear, since the integration is more complicated. But the notion that orthogonal modes of different frequency have different numbers of nodes in space holds rather generally. In fact, a given harmonic mode will generally contain more nodes than lower-frequency modes. This is analogous to the statement that each vibrational mode of a string with fixed ends contains one more node than the one below it. This will be important in our discussion in chapter 5.

Electromagnetic Energy and the Variational Principle

Although the harmonic modes in a dielectric medium can be quite complicated, there is a simple way to understand some of their qualitative features. Roughly, a mode tends to concentrate its displacement energy in regions of high dielectric constant, while remaining orthogonal to the modes below it in frequency. This intuitive notion finds expression in the *electromagnetic variational theorem,* which is analogous to the variational method of quantum mechanics. The lowest-frequency mode, for instance, is the field pattern that minimizes the electromagnetic energy functional:

$$E_f(\mathbf{H}) = \frac{1}{2}\frac{(\mathbf{H}, \Theta\mathbf{H})}{(\mathbf{H}, \mathbf{H})}. \tag{19}$$

To verify this claim, we consider how small variations in $\mathbf{H}(\mathbf{r})$ affect the energy functional. Suppose we add a small displacement $\delta\mathbf{H}(\mathbf{r})$ to $\mathbf{H}(\mathbf{r})$. What is the resulting small change δE_f in the energy functional? It should be zero if the energy functional is really at a minimum, just as the ordinary derivative of a function vanishes at an extremum. To find out, evaluate the energy functional at $\mathbf{H} + \delta\mathbf{H}$ and at $\mathbf{H}$, and then take the difference:

$$E_f(\mathbf{H} + \delta\mathbf{H}) = \frac{1}{2}\frac{(\mathbf{H} + \delta\mathbf{H}, \Theta\mathbf{H} + \Theta\delta\mathbf{H})}{(\mathbf{H} + \delta\mathbf{H}, \mathbf{H} + \delta\mathbf{H})}$$

$$E_f(\mathbf{H}) = \frac{1}{2}\frac{(\mathbf{H}, \Theta\mathbf{H})}{(\mathbf{H}, \mathbf{H})} \tag{20}$$

$$\delta E_f(\mathbf{H}) \equiv E_f(\mathbf{H} + \delta\mathbf{H}) - E_f(\mathbf{H}).$$

Using the real, physical part $\mathbf{H}_r$ of $\mathbf{H}$, and ignoring terms higher than first order in $\delta\mathbf{H}$, we can obtain

$$\frac{\delta E_f(\mathbf{H})}{\delta\mathbf{H}_r} = \frac{1}{(\mathbf{H}_r, \mathbf{H}_r)}\left(\Theta\mathbf{H}_r - \left[\frac{(\mathbf{H}_r, \Theta\mathbf{H}_r)}{(\mathbf{H}_r, \mathbf{H}_r)}\right]\mathbf{H}_r\right). \tag{21}$$

Note that the quantity in brackets vanishes if the magnetic field is an eigenvector of Θ. Therefore, E_f is stationary with respect to variations of $\mathbf{H}$ when $\mathbf{H}$ is a harmonic mode, as we would expect at a minimum or maximum. More careful considerations show that the lowest electromagnetic eigenmode $\mathbf{H}_0$ *minimizes* E_f. The next lowest eigenmode will minimize E_f in the subspace orthogonal to $\mathbf{H}_0$, and so on.

In addition to providing a useful characterization of the modes of Θ, the variational theorem can also give us the physical insight hinted at earlier. Substitution of (16) into (19) gives, after use of (8):

$$E_f(\mathbf{H}) = \left(\frac{1}{2(\mathbf{H}, \mathbf{H})}\right)\int d\mathbf{r}\,\frac{1}{\varepsilon}|\nabla \times \mathbf{H}|^2$$

$$= \left(\frac{1}{2(\mathbf{H}, \mathbf{H})}\right)\int d\mathbf{r}\,\frac{1}{\varepsilon}\left|\frac{\omega}{c}\mathbf{D}\right|^2.$$

(22)

From this expression we can see that E_f is minimized when the displacement field $\mathbf{D}$ is concentrated in the regions of high dielectric constant. To minimize E_f, a harmonic mode will therefore tend to concentrate its displacement field in regions of high dielectric, while remaining orthogonal to the modes below it in frequency. This is the heuristic which we alluded to earlier.

In addition to the variational energy functional, two other important energies of an electromagnetic system are the *physical* energies stored in the electric and magnetic fields:

$$E_D = \left(\frac{1}{8\pi}\right)\int d\mathbf{r}\,\frac{1}{\varepsilon(\mathbf{r})}|\mathbf{D}(\mathbf{r})|^2$$

$$E_H = \left(\frac{1}{8\pi}\right)\int d\mathbf{r}\,|\mathbf{H}(\mathbf{r})|^2.$$

(23)

For a harmonic mode, we can show that $E_D = E_H$, so that as time passes the field energy is harmonically exchanged between the displacement and magnetic fields. Although E_f and E_D have a similar form, there is a very important difference. The energy functional has a normalizing term in the denominator, and allows us to characterize the electromagnetic modes independent of the field strength. The physical energy stored in the electric field, on the other hand, is proportional to the square of the field strength. In other words, if we are interested in the physical energies, we cannot just normalize our fields—an overall multiplier will change the energy. But if we are only interested in mode patterns, overall multipliers are irrelevant.

Why Use the Magnetic Field, and Not the Electric?

In the previous sections, we converted the physics of the Maxwell equations into an eigenvalue condition on the harmonic magnetic modes $\mathbf{H}(\mathbf{r})$. The idea was that for a given frequency, we could solve for $\mathbf{H}(\mathbf{r})$ and then determine the $\mathbf{E}(\mathbf{r})$ via equation (8). But we could have equally well tried the alternate approach: solve for the *electric* field by decoupling equations (6) and then determine the magnetic field with

$$\mathbf{H}(\mathbf{r}) = \frac{ic}{\omega} \nabla \times \mathbf{D}(\mathbf{r}). \tag{24}$$

Why didn't we choose this route? Pursuing this approach, the conditions on the displacement field turn out to be

$$\Xi \mathbf{D}\,(\mathbf{r}) \equiv \nabla \times \left(\nabla \times \frac{1}{\varepsilon(\mathbf{r})}\, \mathbf{D}(\mathbf{r}) \right)$$

$$= \left(\frac{\omega}{c} \right)^2 \mathbf{D}(\mathbf{r}), \quad \nabla \cdot \mathbf{D}(\mathbf{r}) = 0. \tag{25}$$

The operator Ξ is slightly different than Θ; the $\varepsilon(\mathbf{r})$ term is shifted over. In fact, if we try to follow the steps in (13), we find that Ξ is not Hermitian because of the misplaced $\varepsilon(\mathbf{r})$. None of the nice results of the previous sections apply to this problem, since they relied on the Hermiticity of Θ. It seems like a mere cosmetic difference that we might overcome in one of two ways, but in the end both ways are unsatisfactory. First, we might try rewriting the first of equation (6) as

$$\frac{1}{\varepsilon(\mathbf{r})} \nabla \times \left(\nabla \times \frac{1}{\varepsilon(\mathbf{r})} \mathbf{D} \right) = \left(\frac{\omega}{c} \right)^2 \frac{1}{\varepsilon(\mathbf{r})} \mathbf{D}. \tag{26}$$

This is now a *generalized* eigenvalue equation $\Xi_1 \mathbf{D} = (\omega/c)^2\, \Xi_2 \mathbf{D}$, and no longer a simple eigenvalue equation $\Xi \mathbf{D} = (\omega/c)^2\, \mathbf{D}$. It is true that in this case, Ξ_1 and Ξ_2 are both Hermitian operators, so the results of the previous sections follow in modified forms. But it is a far more difficult numerical task to solve a generalized eigenvalue problem than an ordinary eigenvalue problem, so this approach is not practical.

A second apparent "quick-fix" is to define a new field $\mathbf{F}(\mathbf{r})$ as

$$\mathbf{F}(\mathbf{r}) = \frac{1}{\sqrt{\varepsilon(\mathbf{r})}}\, \mathbf{D}(\mathbf{r}) \tag{27}$$

so that equation (26) takes the form

$$\frac{1}{\sqrt{\varepsilon(\mathbf{r})}} \nabla \times \left(\nabla \times \frac{1}{\sqrt{\varepsilon(\mathbf{r})}}\, \mathbf{F}(\mathbf{r}) \right) = \left(\frac{\omega}{c} \right)^2 \mathbf{F}(\mathbf{r})\,. \tag{28}$$

This is a simple eigenvalue equation and the differential operator in (28) is indeed Hermitian. However, the new field $\mathbf{F}$ is not transverse. The transversality requirement is of great utility in numerically evaluating eigenvalues and eigenvectors, so the formulation in terms of $\mathbf{F(r)}$ is also not practical. For these reasons, we employ the *magnetic* field in most of our discussions.

Scaling Properties of the Maxwell Equations

One interesting feature of electromagnetism in dielectric media is that there is no fundamental length scale other than the assumption that the system is macroscopic. In atomic physics, the potentials have the fundamental length scale of the Bohr radius, so that configurations differing only in their absolute length scale nevertheless have very different behaviors. For photonic crystals there is no fundamental constant with the dimensions of length, so there is a simple relationship between electromagnetic problems that differ only by a contraction or expansion of all distances.

Suppose, for example, we have an electromagnetic eigenmode $\mathbf{H(r)}$ of frequency ω in a dielectric configuration $\varepsilon(\mathbf{r})$. From our discussion in the previous sections, we recall the master equation (7):

$$\nabla \times \left(\frac{1}{\varepsilon(\mathbf{r})} \nabla \times \mathbf{H(r)} \right) = \left(\frac{\omega}{c} \right)^2 \mathbf{H(r)} . \tag{29}$$

Now suppose we are curious about the harmonic modes in a configuration of dielectric $\varepsilon'(\mathbf{r})$ that is just a compressed or expanded version of $\varepsilon(\mathbf{r})$: $\varepsilon'(\mathbf{r}) = \varepsilon(\mathbf{r}/s)$ for some scale parameter s. We can just make a change of variables in (29), using $\mathbf{r}' = s\mathbf{r}$ and $\nabla' = \nabla/s$:

$$s\nabla' \times \left(\frac{1}{\varepsilon(\mathbf{r}'/s)} s \, \nabla' \times \mathbf{H(r'}/s) \right) = \left(\frac{\omega}{c} \right)^2 \mathbf{H(r'}/s). \tag{30}$$

But $\varepsilon(\mathbf{r}'/s)$ is none other than $\varepsilon'(\mathbf{r}')$. Dividing out the s's shows that

$$\nabla' \times \left(\frac{1}{\varepsilon'(\mathbf{r}')} \nabla' \times \mathbf{H}(\mathbf{r}'/s) \right) = \left(\frac{\omega}{cs} \right)^2 \mathbf{H}(\mathbf{r}'/s). \tag{31}$$

But this is just the master equation again, this time with mode profile $\mathbf{H'(r')}=\mathbf{H(r'}/s)$ and frequency $\omega' = \omega/s$. In words, if we want to know the new mode profile after changing the length scale by a factor s, we just scale the old mode and its frequency by the same factor. The solution of the problem at one length scale determines the solutions at all other length scales.

This simple fact is of considerable practical importance. For example, the microfabrication of complex micron-scale photonic crystals can be quite difficult. But models can be easily made and tested in the microwave regime, at the much larger length scale of centimeters. Our considerations in this section guarantee that the model will have the same electromagnetic properties.

Just as there is no fundamental *length* scale, there is also no fundamental value of the dielectric constant. Suppose we know the harmonic modes of a system with dielectric configuration $\varepsilon(\mathbf{r})$, and we are curious about the modes of a system with a dielectric configuration that differs by a constant factor everywhere, so that $\varepsilon'(\mathbf{r}) = \varepsilon(\mathbf{r})/s^2$. Substituting $s^2\varepsilon'(\mathbf{r})$ for $\varepsilon(\mathbf{r})$ in (29) yields

$$\nabla \times \left(\frac{1}{\varepsilon'(\mathbf{r})}\nabla \times \mathbf{H}(\mathbf{r})\right) = \left(\frac{s\omega}{c}\right)^2 \mathbf{H}(\mathbf{r}). \qquad (32)$$

The harmonic modes of the new system are unchanged, but the frequencies are all scaled by a factor s: $\omega \rightarrow \omega' = s\omega$. If we multiply the dielectric constant everywhere by a factor of $\frac{1}{4}$, the mode patterns are unchanged but their frequencies double.

Electrodynamics and Quantum Mechanics Compared

As a compact summary of the topics in this chapter, and for the benefit of those readers familiar with quantum mechanics, we now present some similarities between our formulation of electrodynamics in dielectric media and the quantum mechanics of noninteracting electrons (see table 1). This analogy is developed further in appendix A.

In both cases, we decompose the fields into harmonic modes that oscillate with a phase factor $e^{i\omega t}$. In quantum mechanics, the wave function is a complex scalar field. In electrodynamics, the magnetic field is a real vector field and the complex exponential is just a mathematical convenience.

In both cases, the modes of the system are determined by a Hermitian eigenvalue equation. In quantum mechanics, the frequency ω is related to the eigenvalue via $E = h\omega/2\pi$, which is meaningful only up to an overall constant V_0.[4] In electrodynamics, the eigenvalue is proportional to the frequency squared, and there is no arbitrary additive constant.

---

[4]Here h is a fundamental constant called Planck's constant, with value h $\approx 6.626 \times 10^{-27}$ erg sec.

Table 1 Comparison of quantum mechanics and electrodynamics

Field	$\Psi(\mathbf{r}, t) = \Psi(\mathbf{r})e^{i\omega t}$	$\mathbf{H}(\mathbf{r}, t) = \mathbf{H}(\mathbf{r})e^{i\omega t}$
Eigenvalue problem	$H\Psi = E\Psi$	$\Theta\mathbf{H} = (\omega/c)^2\mathbf{H}$
Hermitian operator	$H = \dfrac{-(h/2\pi)^2\nabla^2}{2m} + V(\mathbf{r})$	$\Theta = \nabla \times \left(\dfrac{1}{\varepsilon(\mathbf{r})}\nabla \times \quad\right)$

One difference we did not discuss, but is apparent from table 1, is that in quantum mechanics, the Hamiltonian is separable if $V(\mathbf{r})$ is separable. For example, if $V(\mathbf{r})$ is just a product of functions $V_x(x)V_y(y)V_z(z)$, then the problem separates into three more manageable problems—one for each direction. In electrodynamics, such a factorization is not possible. The differential operator is Θ, which couples the different directions even if $\varepsilon(\mathbf{r})$ is separable. This makes analytical solutions more difficult. To address most of the phenomena of photonic crystals, we will have to make use of numerical solutions.

In quantum mechanics, the lowest eigenstates typically have amplitude concentrated in regions of low potential, while in electrodynamics the lowest modes have their electrical energy concentrated in regions of high dielectric constant. Both of these statements are made quantitative by a variational theorem.

Finally, in quantum mechanics, there is usually a fundamental length scale that prevents us from relating solutions to potentials which differ by a scale factor. Electrodynamics is free from such a length scale, so the solutions we obtain are easily scaled.

Further Reading

One particularly lucid undergraduate text on electromagnetism is Griffiths (1989). A more advanced and complete treatment of the macroscopic Maxwell equations, including a derivation from their microscopic counterparts, is contained in Jackson (1962). To explore the analogy between our formalism and the Schrdinger equation of quantum mechanics, consult the first few chapters of any introductory quantum mechanics text. In particular, Shankar (1982), Liboff (1992), and Sakurai (1985) develop the properties of the eigenstates of a Hermitian operator with proofs very similar to our own. The first two are undergraduate texts; the third is at the graduate level.

3

Symmetries and Solid-State Electromagnetism

If a dielectric structure has a certain symmetry, then we can catalog the electromagnetic modes of that system using that symmetry. In this chapter, we will investigate what various symmetries of a system tell us about its electromagnetic modes. Translational symmetries (both discrete and continuous) are important since photonic *crystals* are *periodic* dielectrics, and because they provide a natural setting for the discussion of band gaps. Some of the terminology of solid-state physics is appropriate, and will be introduced. We will also investigate rotational, mirror, inversion, and time-reversal symmetries.

Using Symmetries to Classify Electromagnetic Modes

In both classical mechanics and quantum mechanics, we learn the lesson that the *symmetries* of a system allow one to make general statements about that system's behavior. Because of the mathematical analogy we pursued in the last chapter, it is not too surprising that symmetry also helps to determine the properties of electromagnetic systems. We will begin with a concrete example of a symmetry and the conclusion we may draw from it, and will then pass on to a more formal discussion of symmetries in electromagnetism.

Suppose we want to find the modes that are allowed in the two-dimensional metal cavity shown in figure 1. Its shape is somewhat arbitrary, so it would be hard to set up an exact boundary condition and solve the problem explicitly. But the cavity has an important symmetry: if you invert the cavity about its center, you end up with exactly the same cavity shape. So if, somehow, we find that the particular pattern $\mathbf{H}(\mathbf{r})$ is a mode with frequency ω, then the pattern

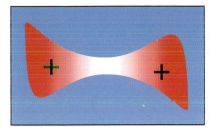
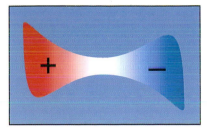

Figure 1 A two-dimensional metallic cavity with inversion symmetry. Red and blue suggest positive and negative fields. On the left, an *even* mode occupies the cavity, for which $\mathbf{H(r)} = \mathbf{H(-r)}$. On the right, an *odd* mode occupies the cavity, for which $\mathbf{H(r)} = -\mathbf{H(-r)}$.

$\mathbf{H(-r)}$ must also be a mode with frequency ω. The cavity cannot distinguish between these two modes, since it cannot tell $\mathbf{r}$ from $-\mathbf{r}$.

Recall from chapter 2 that different modes with the same frequency are said to be *degenerate*. If $\mathbf{H(r)}$ is not degenerate, then since $\mathbf{H(-r)}$ has the same frequency it must be the *same mode*. It must be nothing more than a simple multiple of $\mathbf{H(r)}$: $\mathbf{H(-r)} = \alpha\mathbf{H(r)}$. But what is α? If we invert the system *twice*, picking up another factor of α, then we return to the original function $\mathbf{H(r)}$. Therefore $\alpha^2\mathbf{H(r)} = \mathbf{H(r)}$, and we see that $\alpha = 1$ or -1. A given nondegenerate mode[1] must be one of two types: either it is invariant under inversion, $\mathbf{H(-r)} = \mathbf{H(r)}$, and we call it *even;* or, it becomes its own opposite, $-\mathbf{H(-r)} = \mathbf{H(r)}$, and we call it *odd.* These possibilities are depicted in figure 1. We have classified the modes of the system based on how they respond to one of its symmetry operations.

With this example in mind, we can place this notion in a formal setting. Suppose I is an operator that inverts vectors, so that $I\mathbf{a} = -\mathbf{a}$. To invert a vector *field*, we use an operator O_I that inverts both the vector $\mathbf{f}$ and its argument $\mathbf{r}$: $O_I\mathbf{f(r)} = I\mathbf{f}(I\mathbf{r})$.[2] What is the mathematical expression of the statement that our system has inversion symmetry?

---

[1]This is not true of degenerate modes. But we can always form new modes which *are* even or odd, by taking appropriate linear combinations of degenerate modes.

[2]There is a minor complication here because $\mathbf{H}$ is a pseudovector and $\mathbf{E}$ is a vector, as proven in Jackson (1962). For this reason, $\mathbf{H}$ transforms with a plus sign ($I\mathbf{H} = +\mathbf{H}$), while $\mathbf{E}$ transforms with a minus sign ($I\mathbf{E} = -\mathbf{E}$). That is, $O_I\,\mathbf{H(r)} = +\mathbf{H(-r)}$, and $O_I\,\mathbf{E(r)} = -\mathbf{E(-r)}$. An even mode is defined as one that is invariant under the inversion O_I, which means that an even mode has $\mathbf{H(r)} = \mathbf{H(-r)}$ and $\mathbf{E(r)} = -\mathbf{E(-r)}$. Similarly, an odd mode is defined as one that acquires a minus sign under the inversion O_I, so that $\mathbf{H(r)} = -\mathbf{H(-r)}$ and $\mathbf{E(r)} = \mathbf{E(-r)}$.

Since inversion is a symmetry of our system, it does not matter whether we operate with Θ or we first invert the coordinates, then operate with Θ, and then change them back:

$$\Theta = O_I^{-1} \Theta O_I. \tag{1}$$

This equation can be rearranged as $O_I\Theta - \Theta O_I = 0$. Following this cue, we define the *commutator* $[A, B]$ of two operators A and B just like the commutator in quantum mechanics:

$$[A, B] = AB - BA. \tag{2}$$

Note that the commutator is itself an operator. We have shown that our system is symmetric under inversion only if the inversion operator commutes with Θ; that is, we must have $[O_I, \Theta] = 0$. If we now operate with this commutator on any mode of the system $\mathbf{H}(\mathbf{r})$, we obtain

$$[O_I, \Theta]\mathbf{H} = O_I(\Theta \mathbf{H}) - \Theta(O_I \mathbf{H}) = 0$$

$$\Theta(O_I \mathbf{H}) = O_I(\Theta \mathbf{H}) = \frac{\omega^2}{c^2}(O_I \mathbf{H}). \tag{3}$$

This equation tells us that if $\mathbf{H}$ is a harmonic mode with frequency ω, then $O_I\mathbf{H}$ is *also* a mode with frequency ω. If there is no degeneracy, then there can only be one mode per frequency, so $\mathbf{H}$ and $O_I\mathbf{H}$ can be different only by a multiplicative factor: $O_I\mathbf{H} = \alpha\mathbf{H}$. But this is just the eigenvalue equation for O_I, and we already know that the eigenvalues α must be either 1 or -1. Thus, we can catalog the eigenvectors $\mathbf{H}(\mathbf{r})$ according to whether they are even $(\mathbf{H} \rightarrow +\mathbf{H})$ or odd $(\mathbf{H} \rightarrow -\mathbf{H})$ under the inversion symmetry operation O_I.

What if there *is* degeneracy in the system? Then two modes may have the same frequency, but might not be related by a simple multiplier. Although we will not demonstrate it in detail, we can always form linear combinations of such degenerate modes to make modes which themselves are even or odd.

Generally speaking, whenever two operators commute, one can construct simultaneous eigenfunctions of both operators. This is very convenient, since eigenfunctions and eigenvalues of simple symmetry operators like O_I are easily determined, whereas those for Θ are not. But if Θ commutes with a symmetry operator S, we can construct and catalog the eigenfunctions of Θ using their S-properties. In the case of inversion symmetry, we can classify the Θ-eigenfunctions as either odd or even. We will find this approach useful in later sections when we introduce translational, rotational, and mirror symmetries.

Continuous Translational Symmetry

Another symmetry that a system might have is continuous translation symmetry. Such a system is unchanged if we translate everything through the same distance in a certain direction. Given this information, we can determine the functional form of the system's modes.

A system with translational symmetry is unchanged by a translation through a displacement $\mathbf{d}$. For each $\mathbf{d}$, we can define a translation operator $T_{\mathbf{d}}$ which, when operating on a function $\mathbf{f}(\mathbf{r})$, shifts the argument by $\mathbf{d}$. Suppose our system is translationally invariant; then we have $T_{\mathbf{d}}\, \varepsilon(\mathbf{r}) = \varepsilon(\mathbf{r} + \mathbf{d}) = \varepsilon(\mathbf{r})$, or equivalently, $[T_{\mathbf{d}}, \Theta] = 0$. The modes of Θ can now be classified according to how they behave under $T_{\mathbf{d}}$.

A system with *continuous* translation symmetry in the z-direction is invariant under all of the $T_{\mathbf{d}}$'s for that direction. What sort of function is an eigenfunction of all the $T_{\mathbf{d}}$'s? We can prove that a mode with the functional form e^{ikz} is an eigenfunction of any translation operator in the z-direction:

$$T_d e^{ikz} = e^{ik(z+d)} = (e^{ikd})\, e^{ikz}. \tag{4}$$

The corresponding eigenvalue is e^{ikd}. The modes of our system must be eigenfunctions of all the $T_{\mathbf{d}}$'s, so we know they should have a z-dependence of the functional form e^{ikz}. We can classify them by their particular values for k, the *wave vector*.

A system that has continuous translational symmetry in *all three* directions is free space: $\varepsilon(\mathbf{r}) = 1$. Following a line of argument similar to the one above, we can deduce that the modes must have the form

$$\mathbf{H}_{\mathbf{k}}(\mathbf{r}) = \mathbf{H}_0\, e^{i\,(\mathbf{k}\cdot\mathbf{r})}, \tag{5}$$

where $\mathbf{H}_0$ is any constant vector. These are just plane waves, polarized in the direction of $\mathbf{H}_0$. Imposing the transversality requirement, equation (5) of chapter 2, gives the further restriction $\mathbf{k} \cdot \mathbf{H}_0 = 0$. We can also verify that these plane waves are in fact solutions of the master equation with eigenvalues $(\omega/c)^2 = k^2$. We classify plane waves by specifying $\mathbf{k}$, which specifies how the mode behaves under a translation operation.

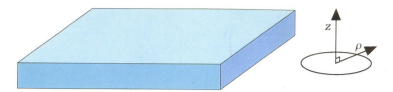

Figure 2   A plane of glass. If the glass extends much further in the x- and y- directions than in the z-direction, we may consider this system to be one-dimensional. The dielectric function $\varepsilon(\mathbf{r})$ varies in the z-direction, but has no dependence on the in-plane coordinate ρ.

Another simple system with continuous translational symmetry is an infinite plane of glass, as shown in figure 2. In this case, the dielectric constant varies in the z-direction, but not in the x- or y-directions: $\varepsilon(\mathbf{r}) = \varepsilon(z)$. The system is invariant under all of the translation operators of the xy-plane. We can catalog the modes using their in-plane wave vector, $\mathbf{k} = k_x\hat{\mathbf{x}} + k_y\hat{\mathbf{y}}$. The x- and y-dependence must once again be like a plane wave (complex exponential):

$$\mathbf{H_k}(\mathbf{r}) = e^{i\,(\mathbf{k}\cdot\rho)}\mathbf{h}(z). \tag{6}$$

As in the above equation, we typically denote a vector that is confined to the xy-plane by ρ. The function $\mathbf{h}(z)$ cannot be determined by this line of reasoning, since the system does not have translational symmetry in that direction. (The transversality condition does imply one restriction on $\mathbf{h}$.)

The reason why the modes must be like equation (6) can also be understood with an intuitive argument. Consider three noncollinear neighboring points at $\mathbf{r}$, $\mathbf{r} + d\mathbf{x}$, and $\mathbf{r} + d\mathbf{y}$, all of which have the same z-value. Due to symmetry, these three points should be treated equally, and should have the same magnetic field amplitude. The only conceivable difference could be the variation in the phase between the points. But once we choose the phase differences between these three points, we set the phase relationships between *all* the points. We have effectively specified k_x and k_y at one point, but they must be universal to the plane. Otherwise we could distinguish different locations in the plane by their phase relationships. Along the z-direction, however, this restriction does not hold. Each plane is at a different distance from the bottom of the glass structure and can conceivably have a different amplitude and phase.

We have seen that we can classify the modes by their values of $\mathbf{k}$.

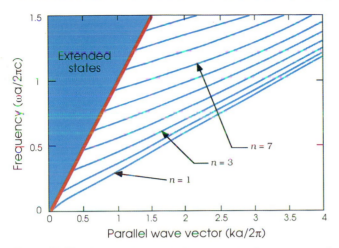

Figure 3 The harmonic mode frequencies for a plane of glass of width a and $\varepsilon = 11.4$. The blue lines correspond to bands of modes which are localized in the glass. The blue region contains a continuum of states which extend into both the glass and the air around it. The red line is the *light line* $\omega = ck$. This plot shows modes of only one polarization; **H** is perpendicular to both the z- and k-directions.

Although we cannot yet say anything about $\mathbf{h}(z)$, we can nevertheless line the modes up (whatever they are) in order of increasing frequency. For a given $\mathbf{k}$, let n stand for a particular mode's place in line of increasing frequency, so that we can identify any mode by its unique name $(\mathbf{k}, n)$. If there is degeneracy, then we might have to include an additional index to name the degenerate modes that have the same n and $\mathbf{k}$.

We call n the *band number*. If there are countably many modes, we can use integers for n, but sometimes n may be a continuous variable. As the n-value grows, so does the frequency of the mode. If we make a plot of wave vector versus mode frequency for the plane of glass, the different bands correspond to different lines that rise uniformly in frequency. This *band structure* is shown in figure 3. We computed it by solving the master equation (7) of chapter 2 numerically.

Let us now be a little more concrete. Consider a plane of glass of width a centered about the origin. For now, we will focus on the particular modes with a wave vector in the y-direction and the magnetic field entirely in the x-direction:

$$\mathbf{H}_{k_y, n}(\mathbf{r}) = e^{i(k_y y)}\phi_n(z)\hat{\mathbf{x}} . \tag{7}$$

Inserting this into master equation (7) of chapter 2, and performing a little algebra, we obtain the condition

$$\nabla \cdot \left(\frac{1}{\varepsilon(\mathbf{r})} \nabla \phi \right) = \frac{d}{dz} \left(\frac{1}{\varepsilon(z)} \frac{d\phi}{dz} \right) = \left(\frac{k_y^2}{\varepsilon(z)} - \frac{\omega^2}{c^2} \right) \phi . \qquad (8)$$

We can catalog the modes of the glass plate according to the character of the fields in the air region. There are two important cases. If $\omega > ck_y$, then the fields will be oscillatory: $\phi(z) \propto (\exp(ik_z z)$. In this case, there is a continuum of states that extend in both the glass and the air region—shown by the blue region of figure 3. On the other hand, if $\omega < ck_y$, then the states decay away to zero in the air regions as $\exp(-\kappa z)$. These states do not belong to a continuum. They come in discrete bands indexed by an integer $n > 0$, which happens to equal to the number of nodes in the dielectric region. For large k_y the frequencies are given by

$$\frac{\omega^2}{c^2} \to \frac{k_y^2}{\varepsilon} + \frac{n^2 \pi^2}{\varepsilon a^2} \qquad (k_y a > n\pi) . \qquad (9)$$

These modes decay ever more rapidly as k_y increases, since $\kappa^2 = k_y^2 - \omega^2/c^2$. The extended states exist above the *light line* $\omega = ck_y$, shown as the red line in figure 3, and the decaying modes are those below the light line.

Discrete Translational Symmetry

Photonic crystals, like the familiar crystals of atoms, do not have continuous translational symmetry; instead, they have *discrete* translational symmetry. That is, they are not invariant under translations of *any* distance—only under distances that are a multiple of some fixed step length. The simplest example of such a system is a structure that is repetitive in one direction, like the configuration in figure 4.

For this system we still have continuous translational symmetry in the x-direction, but now we have *discrete* translational symmetry in the y-direction. The basic step length is the *lattice constant a,* and the basic step vector is called the *primitive lattice vector,* which in this case is $\mathbf{a} = a\hat{\mathbf{y}}$. Because of the symmetry, $\varepsilon(\mathbf{r}) = \varepsilon(\mathbf{r} + \mathbf{a})$. By repeating this translation, we see that $\varepsilon(\mathbf{r}) = \varepsilon(\mathbf{r} + \mathbf{R})$ for any $\mathbf{R}$ that is an integral multiple of $\mathbf{a}$; that is, $\mathbf{R} = l\mathbf{a}$, where l is an integer. The dielectric unit that we consider to be repeated over and over,

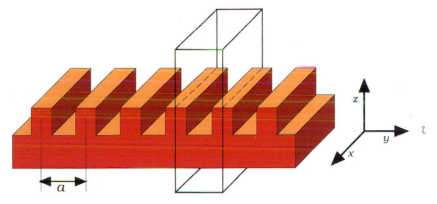

Figure 4 A dielectric configuration with discrete translational symmetry. If we imagine that the system continues forever in the y-direction, then shifting the system by an integral multiple of a in the y-direction leaves it unchanged. The repeated unit of this periodic system is framed with a box. This particular configuration is employed in distributed-feedback lasers, as in Yariv (1985).

highlighted in the figure with a box, is known as the *unit cell*. In this example, the unit cell is an *xz*-slab of dielectric material with width a in the y-direction.

Because of the translational symmetries, Θ must commute with *all* of the translation operators in the x-direction, as well as the translation operators for lattice vectors $\mathbf{R} = la\hat{y}$ in the y-direction. With this knowledge, we can identify the modes of Θ as simultaneous eigenfunctions of both translation operators. As before, these eigenfunctions are plane waves:

$$\begin{aligned}
T_{d\hat{x}} e^{ik_x x} &= e^{ik_x(x+d)} = (e^{ik_x d}) e^{ik_x x} \\
T_{\mathbf{R}} e^{ik_y y} &= e^{ik_y(y+la)} = (e^{ik_y la}) e^{ik_y y}.
\end{aligned} \tag{10}$$

We can begin to classify the modes by specifying k_x and k_y. However, not all values of k_y yield different eigenvalues. Consider two modes, one with wave vector k_y and the other with wave vector $k_y + 2\pi/a$. A quick insertion into (10) shows that they have the same $T_{\mathbf{R}}$-eigenvalues. In fact, all of the modes with wave vectors of the form $k_y + m(2\pi/a)$, where m is an integer, form a degenerate set; they all have the same $T_{\mathbf{R}}$-eigenvalue of $e^{i(k_y la)}$. Augmenting k_y by an integral multiple of $b = 2\pi/a$ leaves the state unchanged. We call $\mathbf{b} = b\hat{y}$ y the primitive *reciprocal* lattice vector.

Since any linear combination of these degenerate eigenfunctions is itself an eigenfunction with the same eigenvalue, we can take linear combinations of our original modes to put them in the form

$$\mathbf{H}_{k_x,\,k_y}(\mathbf{r}) = e^{ik_x x} \sum_m \mathbf{c}_{k_y,\,m}(z)\, e^{i(k_y+mb)y}$$

$$= e^{ik_x x} \cdot e^{ik_y y} \cdot \sum_m \mathbf{c}_{k_y,\,m}(z) e^{imby} \tag{11}$$

$$= e^{ik_x x} \cdot e^{ik_y y} \cdot \mathbf{u}_{k_y}(y,z),$$

where the $\mathbf{c}$'s are expansion coefficients to be determined by explicit solution, and $\mathbf{u}(y, z)$ is (by construction) a periodic function in y. By inspection of equation (11), we can verify that $\mathbf{u}(y + la, z) = \mathbf{u}(y, z)$.

The discrete periodicity in the y-direction leads to a y-dependence for $\mathbf{H}$ that is simply the product of a plane wave with a y-periodic function. We can think of it as a plane wave, as it would be in free space, but modulated by a periodic function because of the periodic lattice:

$$\mathbf{H}(\ldots, y, \ldots) \propto e^{ik_y y} \cdot \mathbf{u}_{k_y}(y, \ldots). \tag{12}$$

This result is commonly known as *Bloch's theorem*. In solid-state physics, the form of (12) is known as a *Bloch state* (as in Kittel 1986), and in mechanics as a *Floquet mode* (as in Mathews and Walker 1964). We will use the former name.

One key fact about Bloch states is that the Bloch state with wave vector k_y and the Bloch state with wave vector $k_y + mb$ are identical. The k_y's that differ by integral multiples of $b = 2\pi/a$ are not different from a physical point of view. Thus the mode frequencies must also be periodic in k_y: $\omega(k_y) = \omega(k_y + mb)$. In fact, we need only consider k_y to exist in the range $-\pi/a < k_y \le \pi/a$. This region of important, nonredundant values of k_y is called the *Brillouin zone*. Readers unfamiliar with the notion of a reciprocal lattice or a Brillouin zone might find appendix B a useful introduction to that material.

We digress briefly to make analogous arguments that apply when the dielectric is periodic in three dimensions; here we skip the details and summarize the results. In this case the dielectric is invariant under translations through a multitude of lattice vectors $\mathbf{R}$ in three dimensions. Any one of these lattice vectors can be written as a particular combination of three primitive lattice vectors $(\mathbf{a}_1, \mathbf{a}_2, \mathbf{a}_3)$ that are said to "span" the space of lattice vectors. In other words, every $\mathbf{R} = l\mathbf{a}_1 + m\mathbf{a}_2 + n\mathbf{a}_3$ for some integers l, m, and n. As explained in appendix B, the vectors $(\mathbf{a}_1, \mathbf{a}_2, \mathbf{a}_3)$ give rise to three primitive *reciprocal* lattice vectors $(\mathbf{b}_1, \mathbf{b}_2, \mathbf{b}_3)$ defined so that $\mathbf{a}_i \cdot \mathbf{b}_j = 2\pi\delta_{ij}$. These reciprocal vectors form a lattice of their own which is inhabited by wave vectors.

The modes of a three-dimensional periodic system are Bloch states that can be labeled by $\mathbf{k} = k_1\mathbf{b}_1 + k_2\mathbf{b}_2 + k_3\mathbf{b}_3$, where $\mathbf{k}$ lies in the Brillouin zone. For example, for a crystal in which the unit cell is a rectangular box, the Brillouin zone is given by $-|\mathbf{b}_i|/2 < |k_i| \leq |\mathbf{b}_i|/2$. Each value of the wave vector $\mathbf{k}$ inside the Brillouin zone identifies an eigenstate of Θ with frequency $\omega(\mathbf{k})$ and an eigenvector $\mathbf{H}_k$ of the form

$$\mathbf{H}_k(\mathbf{r}) = e^{i(\mathbf{k}\cdot\mathbf{r})}\mathbf{u}_k(\mathbf{r}), \tag{13}$$

where $\mathbf{u}_k(\mathbf{r})$ is a periodic function on the lattice: $\mathbf{u}_k(\mathbf{r}) = \mathbf{u}_k(\mathbf{r} + \mathbf{R})$ for all lattice vectors $\mathbf{R}$.

Photonic Band Structures

From very general symmetry principles, we have just suggested that the electromagnetic modes of a photonic crystal with discrete periodicity in three dimensions can be written as Bloch states, as in equation (13). All of the information about such a mode is given by the wave vector $\mathbf{k}$ and the periodic function $\mathbf{u}_k(\mathbf{r})$. To solve for $\mathbf{u}_k(\mathbf{r})$, we insert the Bloch state into the master equation (7) of chapter 2:

$$\Theta\,\mathbf{H}_k = (\omega(\mathbf{k})/c)^2\mathbf{H}_k$$

$$\nabla \times \left(\frac{1}{\varepsilon(\mathbf{r})}\nabla \times e^{i(\mathbf{k}\cdot\mathbf{r})}\mathbf{u}_k(\mathbf{r})\right) = (\omega(\mathbf{k})/c)^2(e^{i(\mathbf{k}\cdot\mathbf{r})}\mathbf{u}_k(\mathbf{r}))$$

$$(i\mathbf{k} + \nabla) \times \left(\frac{1}{\varepsilon(\mathbf{r})}(i\mathbf{k} + \nabla) \times \mathbf{u}_k(\mathbf{r})\right) = (\omega(\mathbf{k})/c)^2\mathbf{u}_k(\mathbf{r}) \tag{14}$$

$$\Theta_k\mathbf{u}_k(\mathbf{r}) = (\omega(\mathbf{k})/c)^2\mathbf{u}_k(\mathbf{r}).$$

Here we have defined $\Theta_{\mathbf{k}}$ as a new Hermitian differential operator that appears in this substitution and depends on $\mathbf{k}$:

$$\Theta_{\mathbf{k}} \equiv (i\mathbf{k} + \nabla) \times \left(\frac{1}{\varepsilon(\mathbf{r})} (i\mathbf{k} + \nabla \times) \right). \tag{15}$$

The function $\mathbf{u}$, and therefore the mode profiles, are determined by the eigenvalue problem in the fourth equation of (14), subject to the condition

$$\mathbf{u}_{\mathbf{k}}(\mathbf{r}) = \mathbf{u}_{\mathbf{k}}(\mathbf{r} + \mathbf{R}). \tag{16}$$

Because of this periodic boundary condition, we can regard the eigenvalue problem as being restricted to a single unit cell of the photonic crystal. As you may remember from "electron-in-a-box" problems in quantum mechanics, restricting an eigenvalue problem to a finite volume leads to a discrete spectrum of eigenvalues. We can expect to find, for each value of $\mathbf{k}$, an infinite set of modes with discretely spaced frequencies, which we can then label with the band index n.

Since $\mathbf{k}$ enters only as a parameter in $\Theta_{\mathbf{k}}$, we expect the frequency of each band, for given $\mathbf{k}$, to vary continuously as $\mathbf{k}$ varies. In this way we arrive at the description of the modes of a photonic crystal. They are a family of continuous functions, $\omega_n(\mathbf{k})$, indexed in order of increasing frequency by the band number. The information contained in these functions is called the *band structure* of the photonic crystal. Studying the band structure of a crystal supplies us with most of the information we need to predict its optical properties, as we will see.

For a given photonic crystal $\varepsilon(\mathbf{r})$, how can we calculate the band structure functions $\omega_n(\mathbf{k})$? Powerful computational techniques are available for the task, but we will not discuss them extensively. The focus of this text is on concepts and results, not on the numerical studies of the equations. A brief outline of the technique that was used to generate the band structures in this text is in appendix D. In essence, the technique relies on the fact that the last equation of (14) is a standard eigenvalue equation that is readily soluble by an iterative minimization technique for each value of $\mathbf{k}$.

Rotational Symmetry and the Irreducible Brillouin Zone

Photonic crystals might have symmetries other than discrete translations. A given crystal might also be left invariant after a rotation, a mirror reflection, or an inversion is performed. For now, we examine the conclusions we can draw about the modes of a system with *rotational* symmetry.

Suppose the operator $\Re(\hat{\mathbf{n}}, \alpha)$ rotates vectors by an angle α about the $\hat{\mathbf{n}}$-axis. Abbreviate $\Re(\hat{\mathbf{n}}, \alpha)$ by $\Re$. To rotate a vector field $\mathbf{f}(\mathbf{r})$, we take the vector $\mathbf{f}$ and rotate it with $\Re$ to give $\mathbf{f}' = \Re\mathbf{f}$. We also rotate the argument $\mathbf{r}$ of the vector field: $\mathbf{r}' = \Re^{-1}\mathbf{r}$. Therefore $\mathbf{f}'(\mathbf{r}') = \Re\mathbf{f}(\mathbf{r}') = \Re\mathbf{f}(\Re^{-1}\mathbf{r})$. Accordingly, we define the vector field rotator $O_\Re$ as

$$O_\Re \cdot f(\mathbf{r}) = \Re f(\Re^{-1}\mathbf{r}). \tag{17}$$

If rotation by $\Re$ leaves the system invariant, then we conclude (as before) that $[\Theta, O_\Re] = 0$. Therefore, we may carry out the following manipulation:

$$\Theta(O_\Re\,\mathbf{H}_{\mathbf{k}n} = O_\Re(\Theta\,\mathbf{H}_{\mathbf{k}n}) = \left(\frac{\omega_n(\mathbf{k})}{c}\right)^2 (O_\Re\,\mathbf{H}_{\mathbf{k}n}). \tag{18}$$

We see that $O_\Re\mathbf{H}_{\mathbf{k}n}$ also satisfies the master equation, with the same eigenvalue as $\mathbf{H}_{\mathbf{k}n}$. This means that the rotated mode is itself an allowed mode, with the same frequency. We can further prove that the state $O_\Re\mathbf{H}_{\mathbf{k}n}$ is none other than the Bloch state with wave vector $\Re\mathbf{k}$. To do this, we must show that $O_\Re\mathbf{H}_{\mathbf{k}n}$ is an eigenfunction of the translation operator $T_\mathbf{R}$ with eigenvalue $e^{i\Re\mathbf{k}\cdot\mathbf{R}}$, where $\mathbf{R}$ is a lattice vector. We can do just that, using the fact that Θ and $O_\Re$ commute:

$$\begin{aligned}
T_\mathbf{R}(O_\Re\,\mathbf{H}_{\mathbf{k}n}) &= O_\Re(T_{\Re^{-1}\mathbf{R}}\mathbf{H}_{\mathbf{k}n}) \\
&= O_\Re(e^{i(\mathbf{k}\cdot\Re^{-1}\mathbf{R})}\mathbf{H}_{\mathbf{k}n}) \\
&= e^{i(\mathbf{k}\cdot\Re^{-1}\mathbf{R})}(O_\Re\mathbf{H}_{\mathbf{k}n}) \\
&= e^{i(\Re\mathbf{k}\cdot\mathbf{R})}(O_\Re\mathbf{H}_{\mathbf{k}n}).
\end{aligned} \tag{19}$$

Since $O_\Re\mathbf{H}_{\mathbf{k}n}$ is the Bloch state with wave vector $\Re\mathbf{k}$ and has the same eigenvalue as $\mathbf{H}_{\mathbf{k}n}$, it follows that

$$\omega_n(\Re\mathbf{k}) = \omega_n(\mathbf{k}) \tag{20}$$

We conclude that when there is rotational symmetry in the lattice, the frequency bands $\omega_n(\mathbf{k})$ have additional redundancies within the Brillouin zone. In a similar manner, we can show that whenever a photonic crystal has a rotation, mirror-reflection, or inversion symmetry, the $\omega_n(\mathbf{k})$ functions have that symmetry as well. This particular

Real lattice

Brillouin zone of
reciprocal lattice

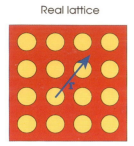
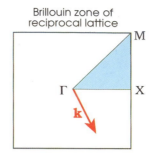

Figure 5 *Left:* A photonic crystal made using a square lattice. An arbitrary vector **r** is shown. *Right:* The Brillouin zone of the square lattice, centered at the origin (Γ). An arbitrary wave vector **k** is shown. The *irreducible zone* is the light blue triangular wedge. The special points at the center, corner, and face are conventionally known as Γ, M, and X.

collection of symmetry operations (rotations, reflections, and inversions) is called the *point group* of the crystal.

Since the $\omega_n(\mathbf{k})$ possess the full symmetry of the point group, we need not consider them at every **k**-point in the Brillouin zone. The smallest region within the Brillouin zone for which the $\omega_n(\mathbf{k})$ are not related by symmetry is called the *irreducible* Brillouin zone. For example, a photonic crystal with the symmetry of a simple square lattice has a square Brillouin zone centered at $\mathbf{k} = 0$, as depicted in figure 5. (See appendix B for a fuller discussion of the reciprocal lattice and the Brillouin zone.) The *irreducible* zone is a triangular wedge with $\frac{1}{8}$ the area of the full Brillouin zone; the rest of the Brillouin zone contains redundant copies of the irreducible zone.

Mirror Symmetry and the Separation of Modes

Mirror reflection symmetry in a photonic crystal deserves special attention. Under certain conditions it allows us to separate the eigenvalue equation for $\Theta_\mathbf{k}$ into two separate equations, one for each field polarization. As we will see, in one case $\mathbf{H}_\mathbf{k}$ is perpendicular to the mirror plane and $\mathbf{E}_\mathbf{k}$ is parallel; while in the other case, $\mathbf{H}_\mathbf{k}$ is in the plane and $\mathbf{E}_\mathbf{k}$ is perpendicular. This simplification is convenient, because it provides immediate information about the mode symmetries and also facilitates the numerical calculation of their frequencies.

To show how this separation of modes comes about, let us turn again to the dielectric system illustrated in figure 4, the notched dielectric. This system is invariant under mirror reflections in the yz- and xz-planes. We focus on reflections M_x in the yz-plane (M_x changes $\hat{\mathbf{x}}$ to $-\hat{\mathbf{x}}$ and leaves $\hat{\mathbf{y}}$ and $\hat{\mathbf{z}}$ alone).[3] In analogy with our rotation operator, we define a mirror reflection operator O_{M_x}, which reflects a vector field by using M_x on both its input and its output:

$$O_{M_x} f(\mathbf{r}) = M_x f(M_x \mathbf{r}). \tag{21}$$

Two applications of the mirror reflection operator restore any system to its original state, so the possible eigenvalues of O_{M_x} are $+1$ and -1. Because the dielectric is symmetric under a mirror reflection in the yz-plane, O_{M_x} commutes with Θ: $[\Theta, O_{M_x}] = 0$. As before, if we operate on $\mathbf{H_k}$ with this commutator we can show that $O_{M_x}\mathbf{H_k}$ is just the Bloch state with the reflected wave vector $M_x\mathbf{k}$:

$$O_{M_x}\mathbf{H_k} = e^{i\phi}\mathbf{H}_{M_x\mathbf{k}}. \tag{22}$$

Here ϕ is an arbitrary phase. This relation does not restrict the reflection properties of $\mathbf{H_k}$ very much, unless $\mathbf{k}$ happens to be pointed in such a way that $M_x\mathbf{k} = \mathbf{k}$. When this is true, (22) becomes an eigenvalue problem and using (21) $\mathbf{H_k}$ must obey

$$O_{M_x}\mathbf{H_k}(\mathbf{r}) = \pm\mathbf{H_k}(\mathbf{r}) = M_x\mathbf{H_k}(M_x\mathbf{r}). \tag{23}$$

Although we will not show it explicitly, the electric field $\mathbf{E_k}$ obeys a similar equation, so that both the electric and magnetic fields must be either *even* or *odd* under the O_{M_x} operation. But $M_x\mathbf{r} = \mathbf{r}$ for any $\mathbf{r}$ in our dielectric. Therefore, since $\mathbf{E}$ transforms like a vector and $\mathbf{H}$ transforms like a pseudovector, the only nonzero field components of an O_{M_x}-*even* mode must be H_x, E_y, and E_z. The *odd* modes are described by the components E_x, H_y, and H_z.

In general, given a reflection M such that $[\Theta, O_M] = 0$, this separation of modes is only possible at $M\mathbf{r} = \mathbf{r}$ for $M\mathbf{k} = \mathbf{k}$. Note from (14) that $\Theta_\mathbf{k}$ and O_M will not commute unless $M\mathbf{k} = \mathbf{k}$. It appears that the separation of polarizations holds only under fairly restricted conditions and is not that useful for three-dimensional photonic crystal analysis. This is not the case for two-dimensional photonic crystals. Two-

---

[3]Note that *any* slice perpendicular to the x-axis is a valid mirror plane for our system. Thus for any $\mathbf{r}$ in the crystal we can always find a plane such that $M_x\mathbf{r} = \mathbf{r}$. This is not true for M_y.

dimensional crystals are periodic in a certain plane, but are uniform along an axis perpendicular to that plane. Calling that axis the z-axis, we know that the operation $\hat{z} \rightarrow -\hat{z}$ is a symmetry of the crystal for any choice of origin. It also follows that $M_z\mathbf{k}_{//} = \mathbf{k}_{//}$ for all wave vectors $\mathbf{k}_{//}$ in the two-dimensional Brillouin zone. Thus the modes of *every* two-dimensional photonic crystal can be classified into two distinct polarizations: either (E_x, E_y, H_z) or (H_x, H_y, E_z). The former, in which the *electric* field is confined to the xy-plane, are called transverse-electric (TE) modes. The latter, in which the *magnetic* field is confined to the xy-plane, are called transverse-magnetic (TM) modes.

Time-Reversal Invariance

We will discuss one more symmetry in detail, and it is of global significance—the *time-reversal* symmetry. If we take the complex conjugate of the master equation for Θ (equation 7 of chapter 2), and use the fact that the eigenvalues are real, we obtain

$$(\Theta \mathbf{H}_{\mathbf{k}n})^* = \frac{\omega_n^2(\mathbf{k})}{c^2}\mathbf{H}_{\mathbf{k}n}{}^*$$

$$\Theta \mathbf{H}_{\mathbf{k}n}{}^* = \frac{\omega_n^2(\mathbf{k})}{c^2}\mathbf{H}_{\mathbf{k}n}{}^*. \tag{24}$$

By this manipulation, we see that $\mathbf{H}_{\mathbf{k}n}^*$ satisfies the same equation as $\mathbf{H}_{\mathbf{k}n}$, with the very same eigenvalue. But from (13) we see that $\mathbf{H}_{\mathbf{k}n}^*$ is just the Bloch state at $(-\mathbf{k}, n)$. It follows that

$$\omega_n(\mathbf{k}) = \omega_n(-\mathbf{k}). \tag{25}$$

The above relation holds for *any* photonic crystal. The frequency bands have inversion symmetry even if the crystal does not. Taking the complex conjugate of $\mathbf{H}_{\mathbf{k}n}$ is equivalent to reversing the sign of time t in the Maxwell equations, as can be verified from equation (4) of chapter 2. For this reason, we say that (25) is a consequence of the *time-reversal* symmetry of the Maxwell equations.

Electrodynamics and Quantum Mechanics Compared Again

As in the previous chapter, we summarize by way of analogy with quantum mechanics. Table 1 compares the system containing an electron

Table 1 Comparison of quantum mechanics and electrodynamics in periodic systems

Discrete translational Symmetry	$V(\mathbf{r}) = V(\mathbf{r} + \mathbf{R})$	$\varepsilon(\mathbf{r}) = \varepsilon(\mathbf{r} + \mathbf{R})$
Commutation relationships	$[H, T_\mathbf{R}] = 0$	$[\Theta, T_\mathbf{R}] = 0$
Bloch's theorem	$\Psi_{\mathbf{k}n}(\mathbf{r}) = u_{\mathbf{k}n}(\mathbf{r})e^{i(\mathbf{k}\cdot\mathbf{r})}$	$\mathbf{H}_{\mathbf{k}n}(\mathbf{r}) = \mathbf{u}_{\mathbf{k}n}(\mathbf{r})e^{i(\mathbf{k}\cdot\mathbf{r})}$

propagating in a periodic potential with the system of electromagnetic modes in a photonic crystal. Appendix A develops this analogy further.

In both cases, the systems have translational symmetry—in quantum mechanics the potential $V(\mathbf{r})$ is periodic, and in the electromagnetic case it is the dielectric function $\varepsilon(\mathbf{r})$. This periodicity implies that the discrete translation operator commutes with the major differential operator of the problem—whether with the Hamiltonian or with Θ.

We can index the eigenstates ($\Psi_{\mathbf{k}n}$ or $\mathbf{H}_{\mathbf{k}n}$) using the translation operator eigenvalues. These can be labeled in terms of the wave vectors and bands in the Brillouin zone. All of the eigenstates can be cast in Bloch form—a periodic function modulated by a plane wave. Qualitatively, the wave undergoes multiple scattering as it moves through the material, but because of the periodicity the scattering is coherent. The field can propagate through the crystal in a coherent manner, as a Bloch wave.

Further Reading

The study of symmetry in the most general context falls under the mathematical subject of group theory or, more specifically, representation theory. Perhaps most useful are texts that apply the formalism of group theory to specific physical disciplines: the first chapter of Harrison (1979) applies the theorems of group theory to solid-state physics, and Hamermesh (1962) does the same for quantum mechanics.

Readers completely unfamiliar with concepts like the reciprocal lattice, the Brillouin zone, or Bloch's theorem might find it useful to consult the first few chapters of Kittel (1986). There, the concepts are introduced where they find common use—in conventional solid-state physics. Additionally, appendix B of this text contains a brief introduction to the reciprocal lattice and the Brillouin zone.

4

The Traditional Multilayer Film:
A One-Dimensional
Photonic Crystal

We begin our study of photonic crystals with the simplest possible case: a one-dimensional system. To understand the propagation of light through a one-dimensional photonic crystal, we apply the principles of electromagnetism and symmetry that we developed in the previous chapters. Even in this simple system we can see the important features of photonic crystals in general—specifically, photonic band gaps and localized modes at defects. Although the optical properties of dielectric layers may be familiar, by casting the discussion in the language of band structures and band gaps we will prepare for the more complicated two- and three- dimensional systems that lie ahead.

The Multilayer Film

The simplest possible photonic crystal, shown in figure 1, consists of alternating layers of material with different dielectric constants. This arrangement is not a very new idea—the optical properties of such *multilayer films* have been widely studied. As we will see, this photonic crystal can act as a perfect mirror for light with a frequency within a sharply-defined gap, and can localize light modes if there are any defects in its structure. This arrangement is commonly used in dielectric mirrors and optical filters (see, for example, Hecht and Zajac 1974).

The traditional approach to an understanding of this system is to allow a plane wave to propagate through the material and to consider the multiple reflections that take place at each interface. In this chapter, we will use a different approach—the analysis of band structures—

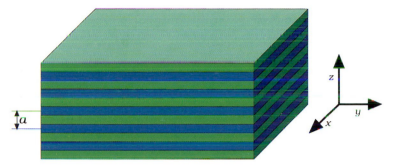

Figure 1 The multilayer film—a one-dimensional photonic crystal. The term "one-dimensional" refers to the fact that the dielectric is only periodic in one direction. (We imagine that the film extends indefinitely in the z-direction.) It consists of alternating layers of materials (blue and green) with different dielectric constants, spaced by a distance a.

which is easily generalized to the more complex two- and three-dimensional photonic crystals.

We begin in the spirit of the previous chapter. By applying symmetry arguments, we can describe the electromagnetic modes sustainable by the crystal. The material is periodic in the z-direction, and homogeneous in the xy-plane. As we saw in the previous chapter, this allows us to index the modes using $\mathbf{k}_{//}$, k_z, and n: the wave vector in the plane, the wave vector in the z-direction, and the band number. The wave vectors tell how the phase of the mode varies with position, and the band number increases with frequency. We can write the modes in the Bloch form:

$$\mathbf{H}_{n,\,k_z\,\mathbf{k}_{//}}(\mathbf{r}) = e^{i\mathbf{k}_{//}\cdot\rho}e^{ik_z z}\mathbf{u}_{n,\,k_z\,\mathbf{k}_{//}}(z). \tag{1}$$

Here, $\mathbf{u}(z)$ is a z-periodic function, so that $\mathbf{u}(z) = \mathbf{u}(z + R)$ whenever R is an integral multiple of a, the layer spacing. The crystal has continuous translational symmetry in the xy-plane, so the wave vector $\mathbf{k}_{//}$ can assume any value. However, we restrict k_z to a finite interval, the one-dimensional Brillouin zone, because the crystal has discrete translational symmetry in the z-direction. Using the prescriptions of the previous chapter, if the primitive lattice vector is $a\hat{\mathbf{z}}$, then the primitive reciprocal lattice vector is $(2\pi/a)\hat{\mathbf{z}}$ and the Brillouin zone is $-\pi/a < k_z \leq \pi/a$.

The Physical Origin of Photonic Band Gaps

For now, consider light that happens to propagate entirely in the z-direction, crossing the sheets of dielectric at normal incidence. In this case, $\mathbf{k}_{//} = 0$, so only the wave vector component k_z is important. Without possibility of confusion, we can abbreviate k_z by k.

In figure 2, we plot $\omega_n(k)$ for three different multilayer films. In the left plot, all of the strips have the same dielectric constant, so the medium is completely homogeneous. The center plot is for a structure with alternating dielectric constants of 13 and 12, and the right-hand plot is for a structure with a much higher dielectric contrast of 13 to 1.[1]

The leftmost plot is for a uniform dielectric medium, to which we have artificially assigned a periodicity of a. But we already know that in a uniform medium, the speed of light is reduced by the index of refraction. The frequency spectrum is just the *light-line* given by

$$\omega(k) = \frac{ck}{\sqrt{\varepsilon}}. \tag{2}$$

Because we have insisted that k repeat itself outside the Brillouin zone, the lines fold back into the zone when they reach the edges. The center plot, which is for a nearly-uniform medium, looks like the light-lines with one important difference. There is a gap in frequency between the upper and lower branches of the lines—a frequency gap in which no mode, regardless of k, can exist in the crystal. We call such a gap a *photonic band gap*. As we can see on the right, as the dielectric contrast is increased, the gap widens considerably.

We will devote a considerable amount of attention to photonic band gaps, and with good reason. Most of the promising applications of two- and three-dimensional photonic crystals to date hinge on the location and width of photonic band gaps. For example, a crystal with a band gap might make a very good, narrow-band filter, by rejecting all (and only) frequencies in the gap. A resonant cavity, carved out of a photonic crystal, would have perfectly reflecting walls for frequencies in the gap.

The natural question arises: Why does the photonic band gap appear? We can understand the gap's physical origin by considering the

---

[1]We use these particular values because the static dielectric constant of gallium arsenide (GaAs) is about 13, and for gallium aluminum arsenide (GaAlAs) it is about 12, as reported in Sze (1981). These materials are commonly used in devices. Air has a dielectric constant $\varepsilon = 1$.

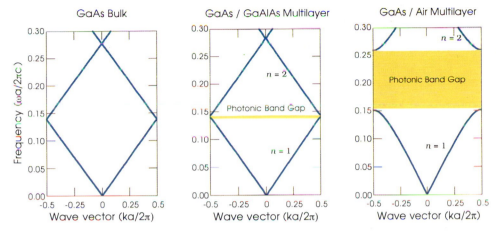

Figure 2 The photonic band structures for on-axis propagation, shown for three differ-ent multilayer films, all of which have layers of width 0.5*a*. *Left:* each layer has the same dielectric constant $\varepsilon = 13$. *Center:* layers alternate between $\varepsilon = 13$ and $\varepsilon = 12$. *Right:* layers alternate between $\varepsilon = 13$ and $\varepsilon = 1$.

electric field mode profiles for the states immediately *above* and *below* the gap. The gap between bands $n = 1$ and $n = 2$ occurs at the edge of the Brillouin zone, at $k = \pi/a$. For now, we focus on the band structure in the center panel of figure 2, corresponding to the configu-ration that is a small perturbation of the uniform system. For $k = \pi/a$, the modes are standing waves with a wavelength of $2a$, twice the crystal's lattice constant.

There are two ways to center a standing wave of this type. We can position its nodes in each low-ε layer, as in figure 3a, or in each high-ε layer, as in figure 3b. Any other position would violate the symmetry of the unit cell about its center.

But in our study of the electromagnetic variational theorem in chapter 2, we found that the low-frequency modes concentrate their energy in the high-ε regions, and the high-frequency modes concen-trate their energy in the low-ε regions. With this in mind, it is under-standable why there is a frequency difference between the two cases. The mode just *under* the gap has its power concentrated in the $\varepsilon = 13$ regions as shown in figure 3c, giving it a lower frequency. Meanwhile, the mode just above the gap has most of its power in lower $\varepsilon = 12$ regions as shown in figure 3d, so its frequency is raised a bit.

The bands above and below the gap can be distinguished by where

(a) E-field for mode at top of band 1

(b) E-field for mode at bottom of band 2

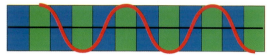

(c) Local power in E-field, top of band 1

(d) Local power in E-field, bottom of band 2

Figure 3 Schematic illustration of the modes associated with the lowest band gap of the center panel of figure 2. (a) Electric field of band 1; (b) electric field of band 2; (c) local energy of band 1; (d) local energy of band 2. In the pictures of the multilayer film, the blue region is the layer of higher dielectric constant ($\varepsilon = 13$).

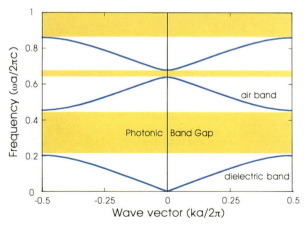

Figure 4 The photonic band structure of a multilayer film with lattice constant a and alternating layers of different widths. The width of the $\varepsilon = 13$ layer is $0.2a$, and the width of the $\varepsilon = 1$ layer is $0.8a$.

the power of their modes lies—in the high-ε regions, or in the low-ε regions. Often the low-ε regions are air regions. For this reason, it is convenient to refer to the band *above* a photonic band gap as the "air band," and the band *below* a gap as the "dielectric band." The situation is analogous to the electronic band structure of semiconductors, in which the "conduction band" and the "valence band" surround the fundamental gap.

Our heuristic, based on the variational theorem, can be extended to describe the configuration with a large dielectric contrast. In this case, we find that the fields for *both* bands are primarily concentrated in the high-ε layers, but in different ways—the bottom band being more concentrated than the top. The gap arises from this difference in field energy location. Consequently, we will still refer to the upper band as the air band, and the lower as the dielectric band.

We conclude this section with the observation that in one dimension, a gap occurs between *every* set of bands, at either the Brillouin zone's edge or its center. This is illustrated for the band structure of a multilayer film in figure 4. Finally, we note that band gaps always appear in one-dimensional photonic crystal for *any* dielectric contrast. The smaller the contrast, the smaller the gaps, but the gaps open up as soon as $\varepsilon_1/\varepsilon_2 \neq 1$.

Evanescent Modes in Photonic Band Gaps

The key observation of the previous section was that the periodicity of the crystal induced a gap into its band structure. No electromagnetic modes are allowed in the gap. But if this is indeed the case, what happens when we send a light wave (with frequency in the photonic band gap) onto the face of the crystal from outside? No purely real wave vector exists for any mode at that frequency. Instead, the wave vector is complex.

The wave amplitude decays exponentially into the crystal. When we say that there are no states in the photonic band gap, we mean that there are no *extended* states like the mode given by equation (1). Instead, the modes are *evanescent,* decaying exponentially:

$$\mathbf{H}(\mathbf{r}) = e^{ikz}\mathbf{u}(z)e^{-\kappa z}. \tag{3}$$

They are just like the modes we constructed in equation (1), but with a complex wave vector $k + i\kappa$. The imaginary component of the wave vector causes the decay on a length scale of $1/\kappa$.

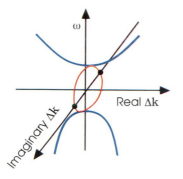

Figure 5 Schematic illustration of the complex band structure of the multilayer film. The upper and lower blue lines correspond to the bottom of band 2 and the top of band 1, respectively. The evanescent states occur on the red line. The maximum decay occurs roughly at the center of the gap.

We would like to understand how these evanescent modes originate, and what determines κ. This can be accomplished by examining the bands in the immediate vicinity of the gap. Return to the right-hand plot of figure 2. Suppose we try to approximate the second band near the gap by expanding $\omega_2(k)$ in powers of k about the zone edge $k = \pi/a$. Because of time-reversal symmetry, the expansion cannot contain odd powers of k, so to lowest order:

$$\Delta\omega = \omega_2(k) - \omega_2\left(\frac{\pi}{a}\right) = \alpha\left(k - \frac{\pi}{a}\right)^2 = \alpha\,(\Delta k)^2. \tag{4}$$

Now we can see where the complex wave vector originates. For frequencies slightly higher than the top of the gap, $\Delta\omega > 0$. In this case, Δk is purely real, and we are within band 2. However, for $\Delta\omega < 0$, when we are within the gap, Δk is purely imaginary. The states decay exponentially since $\Delta k = i\kappa$. As we traverse the gap, the decay constant κ grows as the frequency reaches the gap's center, then disappears at the lower edge. This behavior is depicted in figure 5.

We should emphasize that although evanescent modes are genuine solutions of the eigenvalue problem, they do *not* satisfy the translational-symmetry boundary condition of the crystal. There is no way to excite them in a perfect crystal of infinite extent. However, a defect or an edge in an otherwise perfect crystal might sustain such a mode. One or more evanescent (exponentially decaying) modes may be compatible with the structure and symmetry of a given crystal defect. In those cases, we can create a localized, evanescent light mode within the photonic band gap. And, as a general rule of thumb, we can localize states near the middle of the gap much more tightly than states near the gap's edge.

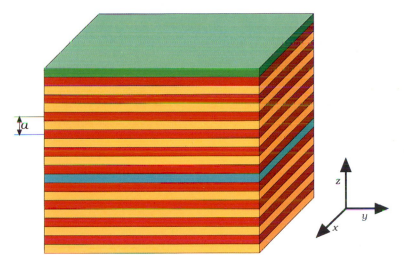

Figure 6 Schematic illustration of possible sites of localized states for a one-dimensional photonic crystal. The states are planar and would be localized near the differently-colored regions, which break the symmetry in the z-direction. We will call a mode at the edge of the crystal (green) a *surface state,* and a mode within the bulk of the crystal (blue) a *defect state.*

Of course, one-dimensional photonic crystals can only localize states near a given *plane*, as shown in figure 6. In the upcoming section, "Localized Modes at Defects," we will discuss the nature of such states when they lie deep within the bulk of a photonic crystal. In certain circumstances, however, an evanescent mode can exist at the face of the crystal. We will also discuss these states, called *surface states*, later in this chapter.

Off-Axis Propagation

So far we have considered the modes of a one-dimensional photonic crystal which happen to have $k_{//} = 0$; that is, modes that propagate only in the z-direction. In this section we will discuss off-axis modes. Figure 7 shows the band structure for modes with $\mathbf{k} = k_y\hat{\mathbf{y}}$ for the one-dimensional photonic crystal described in the caption of figure 4.

The most important difference between on-axis and off-axis propagation is that there are *no band gaps* for off-axis propagation when all possible k_y are considered. This is always the case for a multilayer film, because the off-axis direction contains no periodic dielectric regions to coherently scatter the light and split open a gap.

Another difference involves the *degeneracy* of the bands. For on-axis propagation, the electric field is oriented in the *xy*-plane. We might choose the two basic polarizations as the *x*- and *y*-directions. Since those two modes differ only by a rotational symmetry which the crystal possesses, they must be degenerate. (How could the crystal distinguish the two?)

However, for a mode propagating in some off-axis **k**-direction, this symmetry is broken. The degeneracy is lifted. There are other symmetries; for example, notice that the system is invariant under reflection through the *yz*-plane. For the special case of propagation down the dielectric sheets, in the *y*-direction, we know from the symmetry discussion of chapter 3 that the possible polarizations are in the *x*-direction or in the *yz*-plane. But there is no rotational symmetry relationship between these two bands, so they will generally have different frequencies. All of this information is displayed in figure 7.

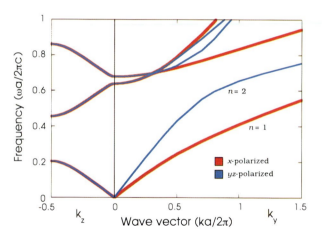

Figure 7 The band structure of a multilayer film. The on-axis bands $(0, 0, k_z)$ are shown on the left side, and an off-axis band structure $(0, k_y, 0)$ is displayed on the right. On-axis, the bands overlap—they are degenerate. Along k_y, the bands split into two distinct polarizations. Red indicates modes polarized so that the electric field points in the *x*-direction, and blue indicates modes polarized in the *yz*-plane. The layered material is the same as the one described in the caption of figure 4.

Although $\omega(\mathbf{k})$ for these two different polarizations have different slopes, both are approximately linear at long wavelengths ($\mathbf{k} \to 0$, $\omega \to 0$). This long-wavelength behavior is characteristic of *all* photonic crystals, regardless of geometry or dimensionality:

$$\omega_\nu(\mathbf{k}) = c_\nu \, (\hat{\mathbf{k}}) \, k. \tag{5}$$

Here ν indexes the two possible polarizations, or equivalently, the first two bands. In general, c_ν will depend on both the direction of $\mathbf{k}$ and the band index.

Why is the dispersion always linear at long wavelengths? At long wavelengths, the electromagnetic wave doesn't probe the fine structure of the crystal lattice. Instead, the light effectively sees a homogeneous dielectric medium; the microscopic bumps of varying ε in the crystal are smoothed out on the light's length scale.

The medium may be anisotropic, having a different average dielectric constant for each direction—these are the effective dielectric constants that we would measure by applying a static field in a capacitance measurement, for example. Typically, we measure three dielectric constants, one for each principal axis of the effective medium. Even if an analytic expression for the effective dielectric constants of a general photonic crystal is not known, they can be calculated numerically.[2]

Returning to the multilayer film, we would like to understand why modes polarized in the x-direction (band 1 in fig. 7) have a lower frequency than modes polarized in the yz-plane (band 2). Once again we use our heuristic: the lower modes concentrate their electrical energy in the high-ε regions. In this case, we focus on the long-wavelength limit of each mode.

The fields for both bands are shown schematically in figure 8. For the x-polarized wave, the displacement fields lie in the high-ε regions. But at long wavelengths, the polarization of band 2 is almost entirely along the z-direction, crossing both the low-ε and the high-ε regions. Continuity forces the field to penetrate the low-ε region, leading to a higher frequency.

---

[2] One useful analytic constraint for the effective dielectric constant of a general photonic crystal is provided by the Weiner bounds, as in Aspnes (1982). Specifically, for a two-material composite, each effective dielectric constant ε_α is bounded by

$$(f_1 \varepsilon_1^{-1} + f_2 \varepsilon_2^{-1})^{-1} \leq \varepsilon_\alpha \leq f_1 \varepsilon_1 + f_2 \varepsilon_2,$$

where f_1 and f_2 are the volume fractions of the materials with dielectric constants ε_1 and ε_2.

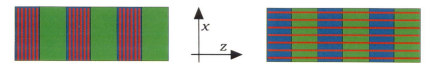

Figure 8 A sketch of the displacement field lines for a long wavelength mode traveling in the *y*-direction (out of the page). In the left figure, the fields are oriented along *x*. In the right figure, the fields are oriented primarily along *z*. The blue regions are the high-ε regions.

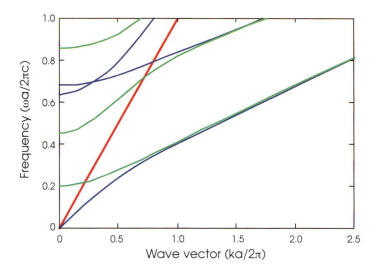

Figure 9 Two superimposed band structures of a multilayer film, showing how the bandwidths vary with k_y. The blue lines refer to bands along $(0, k_y, 0)$, while the green lines are the bands along $(0, k_y, \pi/a)$. Only modes with electric field oriented along the *x*-direction are shown. The red line is the light-line, $\omega = ck_y$. The layered material is the same as the one described in the caption of figure 4.

We can also understand the asymptotic (short-wavelength, large-k) behavior of the modes with a simple argument. In figure 7, note that the range of frequencies spanned by each band (the *bandwidths*) is determined by the difference between frequencies at the zone center ($k = 0$) and zone edge ($k = \pi/a$). At large k_y, the bandwidths decay to zero. This is illustrated in figure 9, which shows the superposition of two band structures. The blue lines represent states along $\mathbf{k} = (0, k_y, 0)$ and the green lines represent states along $\mathbf{k} = (0, k_y, k_z = \pi/a)$.

As we saw for the case of a plane of glass, once the frequency goes below the light line $\omega = ck_y$ the modes decay exponentially into the vacuum region. Therefore, the overlap between modes in neighboring layers of high-ε material goes exponentially to zero. When the coupling between neighboring planes is small, each guides its own mode independent of its neighbors.[3] In this case, the dependence on the on-axis wave vector vanishes, and every mode in the band becomes the frequency of a guided mode, trapped by the high-ε layers.

Localized Modes at Defects

Now that we understand the features of a perfectly periodic system, we can examine systems in which the translational symmetry is broken by a defect. Suppose that the defect consists of a single layer of the one-dimensional photonic crystal that has a different width than the rest. Such a system is shown in figure 10. We no longer have a perfectly periodic lattice, but if we move many wavelengths away from the defect, the modes should behave as before.

For now, we restrict our attention to on-axis propagation and consider a mode with frequency ω in the photonic band gap. There are no extended modes with frequency ω inside the periodic lattice, and introducing the defect will not change that fact. The destruction of periodicity prevents us from describing the modes of the system with wave vector k, but we can still employ our knowledge of the band structure to determine whether a certain frequency will support extended states inside the rest of the crystal. In this way, we can divide up frequency space into regions in which the states are extended and regions in which they are evanescent, as in figure 11.

---

[3] In solid-state physics, the analogous system is the tight-binding model, in the limit of small hopping. See, for example, Harrison (1980).

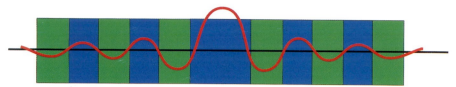

Figure 10 A defect in a multilayer film, formed by enlarging one of the layers of dielectric. Note that this can be considered to be an interface between two perfect multilayer films. Also sketched is the displacement field strength associated with a defect state.

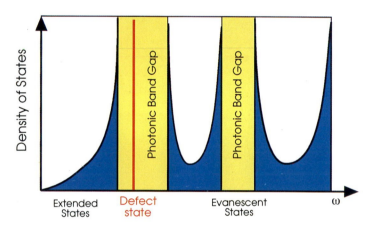

Figure 11 The division of frequency space into extended and evanescent states. In this sketch the *density of states* (the number of allowed modes per unit frequency) is zero in the band gaps of the crystal (yellow). Modes are allowed to exist in these regions only if they are evanescent, and only if the translational symmetry is broken by a defect. Such a mode is shown in red.

Defects may permit *localized* modes to exist, with frequencies inside photonic band gaps. If a mode has a frequency in the gap, then it must exponentially decay once it enters the crystal. The multilayer films on both sides of the defect behave like frequency-specific mirrors. If two such films are oriented parallel to one another, any z-propagating light trapped between them will just bounce back and forth between these two mirrors. And because the distance between the mirrors is of the order the light's wavelength, the modes are *quantized*. The situation bears strong resemblance to the quantum-mechanical problem of a particle in a box (as in Liboff 1992), or the electromagnetic problem of microwaves in a metallic cavity (as in Jackson 1962).

Consider the family of localized states generated by continuously increasing the thickness of the defect layer. The bound mode associated with each member of this family will have a different frequency. As the thickness of the high-ε layer is increased, the frequency will decrease, because the field will be concentrated more and more in a high-ε region. Moreover, the rate of decay will be largest when the frequency is near the center of the gap, as shown in figure 5. States with frequencies in the center of the gap will be most strongly attached to the defect.

The *density of states* of a system is the number of allowed states per unit increase in ω. If a single state is introduced into the photonic band gap, then the density of states of the system in figure 11 is zero in the photonic band gap, except for a single peak associated with the defect. This property is exploited in the band-pass filter known as the *dielectric Fabry-Perot filter*. It is particularly useful at visible-light frequencies, because of the relatively low losses of dielectric materials.

This treatment can be extended to the interface between two multilayer films with different spacings. Localized states can exist as long as the band gaps of the two materials overlap. We can also obtain states that are localized in the z-direction, but propagate along the interface ($k_z = i\kappa$, $\mathbf{k}_{//} \neq 0$).

Surface States

We have seen under what conditions we can localize electromagnetic modes at defects in a multilayer film. In a similar fashion, we can also localize modes at its *surface*. In the previous section, the mode was bound because its frequency was within the photonic band gap of the

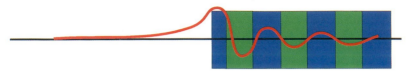

Figure 12 The electric field strength associated with a localized mode at the surface of a multilayer film.

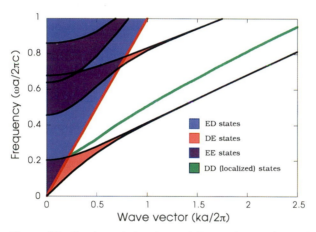

Figure 13 The band structure at the surface of a multilayer film. The shaded regions describe states which are extended in the air region (blue), in the layered material (red), or in both (purple). The green line represents a band of surface states confined at the interface. The layered material is the same as the one described in figure 4. The surface is terminated with a layer of high dielectric with a width 0.1a.

films on both sides. But at the surface, there is only a band gap on one side of the interface—the outside air does not present a band gap.

In this case, light is bound to the surface if its frequency is below the light line. We can think of such a wave as being totally internally reflected (see fig. 12). At a surface we must consider whether the modes are extended or decaying in both the air and the layered material, and we must consider all possibilities for $k_{//}$. The appropriate band structure is shown in figure 13. We divide the phase space of modes into four regions, classified by how they behave in the air and crystal regions. For example, the label "*DE*" means that modes in that region *Decay* in the air region, and are *Extended* in the crystal region.

The *EE* modes extend on both sides of the surface, the *DE* modes decay in the air region and extend into the crystal, and the *ED* modes extend in the air region, but decay inside the crystal. Only if the modes are evanescent on both sides of the surface can we have a surface wave. The region where this is possible is labeled *DD*. In fact, every layered material has surface modes for some termination, a phenomenon we shall discuss again in chapter 6.

Further Reading

Many of the theorems we have developed and the properties we have observed for photonic crystals have analogs in quantum mechanics and solid-state physics. For readers familiar with those fields, appendix A provides a comprehensive listing of these analogs.

The conventional treatment of the multilayer film, including the calculation of absorption and reflection coefficients, can be found in Hecht and Zajac (1974). The use of the multilayer films in optoelectronic devices is widespread in current literature. For example, Fowles (1975) outlines their use in Fabry-Perot filters, and Yeh (1988, p. 337) explains how they are incorporated into distributed feedback lasers.

The details of the computational scheme used to compute band structures can be found in Meade et al. (1993a). Other methods for computing band structures are outlined in Ho et al. (1990) and Sözüer et al. (1992).

5

Two-Dimensional
Photonic Crystals

Now that we have discussed some interesting properties of one-dimensional photonic crystals, in this chapter we will see how the situation changes when the crystal is periodic in two directions and homogeneous in the third. Photonic band gaps appear in the plane of periodicity. For light propagating in this plane, we can separate the modes into two independent polarizations and discuss the band structure of each. As before, we can introduce defects in order to localize light modes, but in addition to planar localization we can achieve linear localization.

Two-Dimensional Bloch States

A two-dimensional photonic crystal is periodic along two of its axes and homogeneous along the third. A typical specimen, consisting of a square lattice of dielectric columns, is shown in figure 1. For certain values of the column spacing, this crystal can have a photonic band gap in the xy-plane. Inside this gap, no extended states are permitted, and incident light is reflected. But although the multilayer film only reflects light at normal incidence, this two-dimensional photonic crystal can reflect light incident from any direction in the plane.

As always, we can use the symmetries of the crystal to characterize its electromagnetic modes. Because the system is homogeneous in the z-direction, we know that the modes must be oscillatory in that direction, with no restrictions on the wave vector k_z. In addition, the system has discrete translational symmetry in the xy-plane. Specifically, $\varepsilon(\rho) = \varepsilon(\rho + \mathbf{R})$, as long as $\mathbf{R}$ is any linear combination of the primitive lattice vectors $a\hat{\mathbf{x}}$ and $a\hat{\mathbf{y}}$. By applying Bloch's theorem, we can focus our attention on the values of $\mathbf{k}_{//}$ that are in the Brillouin zone. As before,

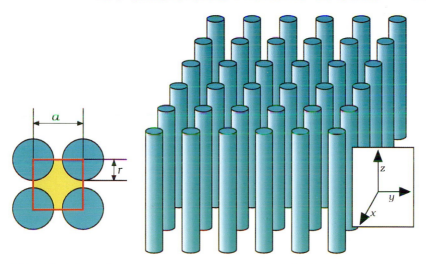

Figure 1 A two-dimensional photonic crystal. This material is a square lattice of dielectric columns, with radius r and dielectric constant ε. The material is homogeneous in the z direction (we imagine the cylinders are very tall), and periodic along x and y with lattice constant a. The left inset shows the square lattice from above, with the unit cell framed in red.

we use the label n (band number) to label the modes in order of increasing frequency.

Indexing the modes of the crystal by k_z, $\mathbf{k}_{//}$ and n, they take the now-familiar form

$$\mathbf{H}_{(n,\, k_z,\, k_{//})}(\mathbf{r}) = e^{i\mathbf{k}_{//}\cdot\rho}e^{ik_z z}\mathbf{u}_{(n,\, k_z,\, k_{//})}(\rho). \tag{1}$$

Here $\mathbf{u}(\rho)$ is a periodic function, $\mathbf{u}(\rho) = \mathbf{u}(\rho + \mathbf{R})$, for all lattice vectors $\mathbf{R}$. The modes of this system look similar to those of the multilayer film that we saw in equation (1) of chapter 4. The key difference is that in the present case, $\mathbf{k}_{//}$ is restricted to the Brillouin zone and k_z is unrestricted. In the multilayer film, the roles of these two vectors were reversed. Also, $\mathbf{u}$ is now periodic in the plane, and not in the z-direction as before.

If $k_z = 0$, so that light is propagating strictly in the xy-plane, then the system is invariant under reflections through the xy-plane. As discussed in chapter 3, this mirror symmetry allows us to classify the modes by separating them into two distinct polarizations. Transverse-electric (TE) modes have **H** normal to the plane, $\mathbf{H} = H(\rho)\hat{\mathbf{z}}$, and **E** in the plane, $\mathbf{E}(\rho) \cdot \hat{\mathbf{z}} = 0$. Transverse-magnetic (TM) modes have just the reverse: $\mathbf{E} = E(\rho)\hat{\mathbf{z}}$ and $\mathbf{H}(\rho) \cdot \hat{\mathbf{z}} = 0$

The band structures for TE and TM modes can be completely different; in particular, there can be photonic band gaps for one and not the other. In the coming sections, we will investigate the TE and TM band structures for two different two-dimensional photonic crystals. The results will provide some useful insights into the appearance of band gaps.

A Square Lattice of Dielectric Columns

Consider light that propagates in the *xy*-plane of a square array of dielectric columns, like the one shown in figure 1. The band structure for a crystal consisting of alumina ($\varepsilon = 8.9$) rods with $r/a = 0.2$ is plotted in figure 2. Both the TE and the TM band structures are shown. Along the horizontal axis (not necessarily to scale), the in-plane wave vector $\mathbf{k}_{//}$ goes along the edge of the irreducible Brillouin zone, from Γ to X to M as shown in the inset to figure 2.

This is the first photonic crystal we have encountered that exhibits a complicated band structure, so we will discuss it in some detail.

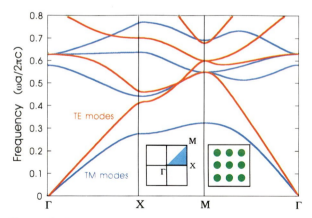

Figure 2 The photonic band structure for a square array of dielectric columns with $r = 0.2a$. The blue bands represent TM modes and the red bands represent TE modes. The left inset shows the Brillouin zone, with the irreducible zone shaded light blue. The right inset shows a cross-sectional view of the dielectric function—the circles ($\varepsilon = 8.9$) are embedded in air ($\varepsilon = 1$).

Specifically, we will describe the nature of the modes when $\mathbf{k}_{//}$ is right at the special symmetry points of the Brillouin zone, and we will investigate the appearance of the band gaps.

The square lattice array has a square Brillouin zone, which is inset in figure 2. The irreducible Brillouin zone is the triangular wedge in the upper-right corner; the rest of the Brillouin zone can be related to this wedge by rotational symmetry. The three special points Γ, X, and M correspond (respectively) to $\mathbf{k}_{//} = 0$, $\mathbf{k}_{//} = \pi/a\hat{\mathbf{x}}$, and $\mathbf{k}_{//} = \pi/a\hat{\mathbf{x}} + \pi/a\hat{\mathbf{y}}$. What do the field profiles of the electromagnetic modes at these points look like?

The field patterns of the TM modes of the first band ("dielectric band") and second band ("air band") are shown in figure 3. For modes at the Γ-point, the fields are the same in each unit cell. The X-point is at the zone edge, so the fields alternate in each unit cell along the direction of the wave vector k_x, forming wave fronts parallel to the y-direction. At M, the phases of the fields alternate in neighboring cells, forming a checkerboard pattern, like a plane wave propagating in the direction $\hat{\mathbf{x}} + \hat{\mathbf{y}}$. The field patterns of the TE modes for the first and second bands are shown in figure 4.

Note that this photonic crystal has a complete band gap for the TM modes (between the first and second bands), but not for the TE modes. We should be able to explain such a significant fact, and we can, by examining the field patterns in figures 3 and 4. The field associated with lowest TM mode (dielectric band) is strongly concentrated in the dielectric regions. This is in sharp contrast to the field pattern of the air band. There, a nodal plane cuts through the dielectric columns, expelling some of the displacement field amplitude from the high-ε region.

As we found in chapter 2, a mode concentrates most of its displacement energy in the high-ε regions in order to lower its frequency. This statement of the variational theorem explains the large splitting between these two bands. The first band has most of its power in the dielectric regions, and has a low frequency; the second has most of its power in the air region, and has a higher frequency. We can quantify this claim. An appropriate measure of the degree of concentration of the displacement fields in the high-ε regions is the *fill factor*, defined as

$$f = \frac{\displaystyle\int_{V_\varepsilon = 8.9} \mathbf{E}^*(\mathbf{r}) \cdot \mathbf{D}(\mathbf{r}) d^3\mathbf{r}}{\displaystyle\int \mathbf{E}^*(\mathbf{r}) \cdot \mathbf{D}(\mathbf{r}) d^3\mathbf{r}}. \tag{2}$$

D-field at Γ-point

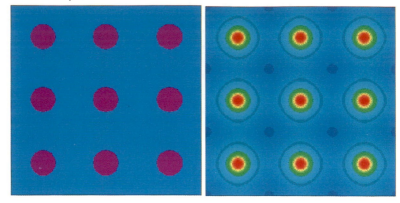

D-field at X-point

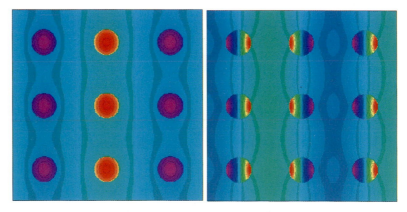

D-field at M-point

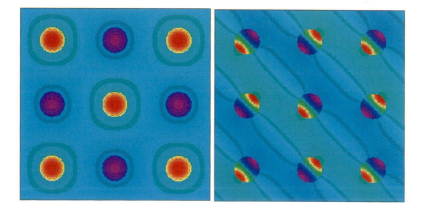

negative ▬▬▬ positive

Figure 3 *Displacement* fields of TM states inside a square array of dielectric ($\varepsilon = 8.9$) columns in air. The color indicates the amplitude of the displacement field, which points in the z-direction. Modes are shown at the Γ point (top), the X point (middle), and the M point (bottom). In each set, band 1 is on the left, and band 2 on the right. The fields of band 2 at the M-point are from one of a pair of degenerate states.

Band 1   Band 2

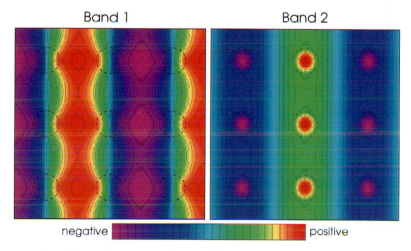

negative   positive

Figure 4 *Magnetic* fields of X-point TE states inside a square array of dielectric ($\varepsilon = 8.9$) columns in air. The column positions are indicated by dotted lines, and the color indicates the amplitude of the magnetic field. The dielectric band is on the left, and the air band on the right. Since **D** is largest along the nodal planes of **H**, the light blue regions are where the displacement energy is concentrated. TE modes have **D** lying in the xy-plane.

Table 1 The fill factors for the lowest two bands of the square lattice of dielectric rods at the X-point of the Brillouin zone

	TM	TE
Dielectric band	0.83	0.23
Air band	0.32	0.09

The fill factor measures the fraction of electrical energy located inside the high-ε regions. Table 1 shows the fill factors for the fields we are considering. The dielectric-band TM mode has a fill factor of 0.83, while the air-band TM mode has a fill factor of only 0.32. This difference in the energy distribution of consecutive modes is responsible for the large TM photonic band gap.

The fill factors for the TE modes do not contrast as strongly. This is reflected in the field configurations for the lowest two bands, shown in figure 4. We have actually plotted the magnetic field **H**, since it is a scalar for TE modes and easy to visualize. But from equation (8) of chapter 2, we know that the displacement field **D** will be largest along the nodal planes of the magnetic field **H**. The displacement field of both modes has a significant amplitude in the air regions, raising the mode frequencies. But in this case there is no choice—there is no continuous pathway between the rods that can contain the field lines of **D**. The field lines must be continuous, so they are forced to penetrate the air regions. This is the origin of the low fill factors, and explains the absence of a band gap for TE modes.

The *vector* nature of the electromagnetic field is central to this argument. The scalar D_z field of the TM modes can be localized within the rods, but the continuous field lines of the TE modes are compelled to penetrate the air regions to connect neighboring rods. As a result, consecutive TE modes cannot exhibit markedly different fill factors, and band gaps do not appear.

A Square Lattice of Dielectric Veins

Another two-dimensional photonic crystal that we will investigate is a square grid of dielectric veins, shown as an inset in figure 5. In a sense, this structure is complementary to the square lattice of dielectric columns, because it is a *connected* structure. The high-ε regions form

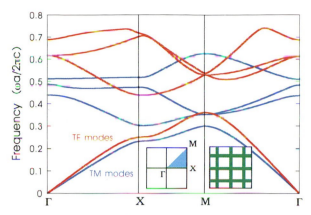

Figure 5 The photonic band structure for the lowest-frequency modes of a square array of dielectric ($\varepsilon = 8.9$) veins in air. The blue lines are TM bands and the red lines are TE bands. The left inset shows the high-symmetry points at the corners of the irreducible Brillouin zone (shaded light blue). The right inset shows a cross-sectional view of the dielectric function.

a continuous path in the *xy*-plane, instead of discrete spots. The complementary nature is reflected in the band structure of figure 5. Here there is a gap in the TE band structure, but not for the TM modes. The opposite was true of the square lattice of dielectric columns.

Again, we turn to the field patterns of the modes in the two lowest bands to understand the appearance of the band gap. The fields are displayed for the TM and TE modes in figures 6 and 7, respectively.

Looking at the TM-field patterns in the dielectric and air bands, we see that both modes are predominately contained within the high-ε regions. The fields of the dielectric band are confined to the dielectric crosses and vertical veins, whereas the fields of the air band are concentrated in the horizontal dielectric veins connecting the square lattice sites. The consecutive modes both manage to concentrate in high-ε regions, thanks to the arrangement of the dielectric veins, so there is no large jump in frequency. This claim is verified by calculating the fill factors for the field configurations. The results are in table 2.

On the other hand, the TE band structure has a photonic band gap between the air and dielectric bands. What is the difference between *this* lattice and the square lattice of dielectric columns, in which no TE gap appeared? In this case, the continuous field lines of the transverse

Band 1   Band 2

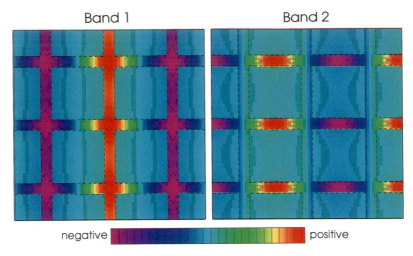

negative   positive

Figure 6 *Displacement* fields of X-point TM modes for a square array of dielectric ($\varepsilon = 8.9$) veins in air. The dotted lines indicate the veins, and the color indicates the amplitude of the displacement field, which is oriented in the z-direction (out of the page). The dielectric band is on the left, and the air band is on the right.

Band 1   Band 2

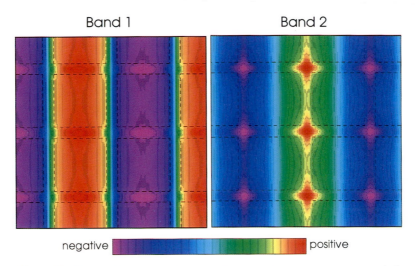

negative   positive

Figure 7 *Magnetic* fields of X-point TE states for a square array of dielectric ($\varepsilon = 8.9$) veins in air. The dotted lines indicate the veins, and the color indicates the amplitude of the magnetic field. The dielectric band is on the left, and the air band is on the right. Recall that **D** is highest near the nodal planes of **H**—the light blue regions.

Table 2 The fill factor for the lowest two bands of the square lattice of veins at the X-point

	TM	TE
Dielectric band	89	83
Air band	77	14

D-field can extend to neighboring lattice sites without ever leaving the high-ε regions. The veins provide high-ε roads for the fields to travel on, and for $n = 1$ the fields stay entirely on them. As proof, figure 7 shows that the lowest band is strongly localized in the vertical dielectric veins.

The **D**-field of the next TE band ($n = 2$) is forced to have a node passing through the vertical high-ε region, to make it orthogonal to the previous band. Some of its energy (indicated by light blue in fig. 7) is thereby forced into the low-ε regions, which corresponds to a sizable jump in frequency. This hypothesis is supported quantitatively by fill factor calculations in table 2. We find a large fill factor for the dielectric band and a small one for the air band; this jump in fill factor between consecutive bands results in the formation of a band gap. In this case, it is the *connectivity* of the lattice that is crucial to the production of TE band gaps.

A Complete Band Gap for All Polarizations

In the previous two sections, we used the field patterns as our guide to understanding which aspects of two-dimensional photonic crystals lead to TM and TE band gaps. By combining our observations, we can design a photonic crystal that has band gaps for *both* polarizations. By adjusting the dimensions of the lattice, we can even arrange for the band gaps to overlap, resulting in a *complete band gap* for all polarizations.

Earlier we found that the isolated high-ε spots of the square lattice of dielectric columns forced consecutive TM modes to have different fill factors (due to the appearance of a node in the higher mode). This, in turn, led to the large TM photonic band gap. The lattice of dielectric veins presented a more spread out distribution of high-ε material, so the fill factors were more uniform.

On the other hand, the connectivity of the veins was the key to achieving gaps in the TE band structure. In the square lattice of dielectric rods, the TE modes were forced to penetrate the low-ε regions, since the field lines had to be continuous. As a result, the fill factors for consecutive modes were both low and not very far apart. This problem disappeared for the lattice of dielectric veins, since the fields could follow the high-ε paths from site to site, and the additional node in the higher mode corresponded to a large frequency jump.

Summarizing our rule of thumb: *TM band gaps are favored in a lattice of isolated high-ε regions, and TE band gaps are favored in a connected lattice.*

It seems impossible to arrange a photonic crystal with both isolated spots and connected regions of dielectric material. The answer is a sort of compromise: we can imagine crystals with high-ε regions that are both practically isolated *and* linked by narrow veins. An example of such a system is the *triangular* lattice of air columns, shown in figure 8.

The idea is to put a triangular lattice of low-ε columns inside a medium with high ε. If the radius of the columns is large enough, the spots between columns look like localized regions of high-ε material, which are connected (through a narrow squeeze between columns) to adjacent spots. This is shown in figure 9.

The band structure for this lattice, shown in figure 10, has photonic band gaps for both the TE and TM polarizations. In fact, for the particular radius $r/a = 0.48$ and dielectric constant $\varepsilon = 13$, these gaps overlap.

The extent of a photonic band gap can be characterized by its frequency width $\Delta\omega$, but this is not a really useful measure. Remember

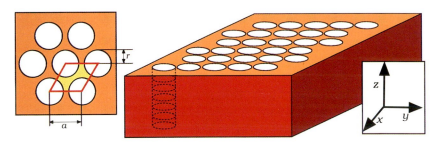

Figure 8 A two-dimensional photonic crystal of air columns in a dielectric substrate. The columns have radius r and dielectric constant $\varepsilon = 1$. The left inset shows a view of the triangular lattice from above, with the unit cell framed in red. It has a lattice constant a.

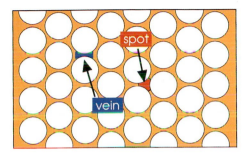

Figure 9 The spots and veins of a triangular lattice. Between the columns are narrow veins, connecting the spots which are surrounded by three columns.

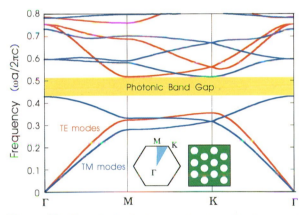

Figure 10 The photonic band structure for the modes of a triangular array of air columns drilled in a dielectric substrate ($\varepsilon = 13$). The blue lines represent TM bands and the red lines represent TE bands. The inset shows the high-symmetry points at the corners of the irreducible Brillouin zone (shaded light blue). Note the complete photonic band gap.

from chapter 2 that all of our results are scalable, and the corresponding band gap in a crystal that is expanded by a factor s would have width $\Delta\omega/s$. A more useful characterization, which is independent of the scale of the crystal, is the *gap-midgap ratio*. Letting ω_0 be the frequency at the middle of the gap, we define the gap-midgap ratio as $\Delta\omega/\omega_0$. If the system is scaled up or down, all of the frequencies scale accordingly, but this quantity stays the same. For the complete photonic band gap in figure 10, the gap-midgap ratio is 0.186.

Out-of-Plane Propagation

Until now we focused exclusively on modes that propagate in the plane of periodicity, so that $k_z = 0$. However, for some applications we must understand the propagation of light in an arbitrary direction. We will investigate the out-of-plane band structure by considering the $k_z > 0$ modes of the triangular lattice of air columns, the lattice which we discussed in the previous section. The out-of-plane band structure for this photonic crystal is shown in figure 11. Many of the qualitative features of the out-of-plane band structure for this crystal are common to all two-dimensional crystals. In fact, these features are just the natural extensions of the corresponding notions in multilayer films, which we developed in the previous chapter.

The first thing to notice about the out-of-plane band structure is that there are no band gaps for propagation in the z-direction. Simply put, this is because the crystal is homogeneous in that direction, so no

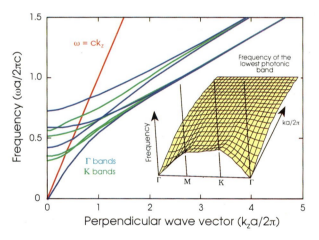

Figure 11 The out-of-plane band structure of the triangular lattice of air columns for the first few bands. The bands which start at Γ, $\omega(\Gamma, k_z)$, are plotted with blue lines, whereas the bands which start at K, $\omega(K, k_z)$, are plotted with green lines. The light line $\omega = ck_z$ (red) separates the modes which are oscillatory ($\omega > ck_z$) in the air regions from those which are evanescent ($\omega < ck_z$) in the air regions. The inset shows the frequency dependence of the lowest band as k_z varies. Note that as k_z increases, the lowest band flattens.

scattering occurs. Remember that it is the multiple scatterings from the regions of different ε that is ultimately responsible for the band structure.

Also note that the bands become flat with increasing k_z. The inset to figure 11 shows the frequency dependence of the lowest band as k_z is varied. When $k_z = 0$, this lowest band spans a broad range of frequencies, as we saw in figure 10. As k_z increases, the lowest band flattens and the bandwidth, the range of allowed frequencies for a given k_z, tends to zero. Since the bandwidth is typically determined by the difference between the frequencies at Γ and K, figure 11 also shows both $\omega(\Gamma, k_z)$ and $\omega(K, k_z)$ for the first few bands. As k_z increases, the bandwidth of each of the bands vanishes. Why is this the case?

There is a simple explanation. For large k_z, the light is trapped by total internal reflection inside the dielectric regions, just as it might be in an optical waveguide. The light modes that are trapped in neighboring high-ε regions have very little overlap, so the modes decouple and the bandwidth shrinks to zero.

This is especially true for modes that have $\omega \ll ck_z$. In this regime, the fields exponentially decay outside the high-ε regions, and the overlap between modes in neighboring high-ε regions vanishes. This behavior is displayed by the bands in figure 11, which have a large dispersion for $\omega > ck_z$, but a small dispersion for $\omega \ll ck_z$.

Localization of Light by Linear Defects

Previously we found two-dimensional photonic crystals with band gaps for in-plane propagation. No modes were allowed with frequencies inside the gap; the density of states, the number of possible modes per unit frequency, is zero within the photonic band gap. By perturbing a single lattice site, we can permit a single (localized) mode or set of closely spaced modes that have frequencies in the gap. When investigating the multilayer film, we found that we could localize light near a perturbed plane in this fashion.

We have several options in two dimensions. As depicted in figure 12, we can remove a single column from the crystal, or replace it with another whose size, shape, or dielectric constant is different than the original. Perturbing just one site ruins the translational symmetry of the lattice, so (strictly speaking) we can no longer classify the modes by an in-plane wave vector. But the mirror-reflection symmetry is still intact for $k_z = 0$, so if we restrict our attention to in-plane propagation,

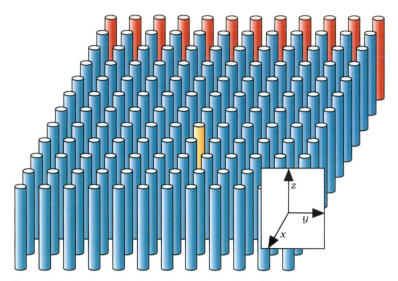

Figure 12 Schematic illustration of possible sites of planar (*yz*-plane) and linear (*z*-direction) defects. Perturbing the line of columns at the surface (red) might allow a localized surface state to exist. Perturbing one column in the bulk of the crystal (orange) might allow a localized defect state to exist.

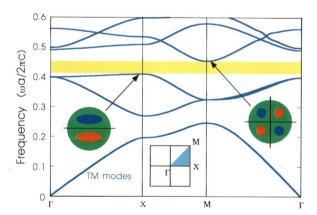

Figure 13 The TM modes of a square array of dielectric (ε = 8.9) columns in air, with $r = 0.38a$. The lower inset shows the irreducible Brillouin zone (shaded light blue). The other two insets suggest the field patterns of the modes, within each column—red for positive field, and blue for negative. The left inset shows the π-like pattern for band 3; the right shows the δ-like pattern for the bottom of band 4.

the TE and TM modes still decouple. That is, we can discuss the band structures for the two polarizations independently, as before.

Removing one column may introduce a peak into the crystal's density of states. If the peak happens to be located in the photonic band gap, then the defect-induced state must be evanescent—the defect mode cannot penetrate the rest of the crystal, since it has a frequency in the band gap. The analysis of chapter 4 is easily generalized to the case of two dimensions, allowing us to conclude that the defect modes decay *exponentially* away from the defects. They are localized in the *xy*-plane, but extend in the *z*-direction.

We reiterate the simple explanation for the localizing power of defects: the photonic crystal, because of its band gap, reflects light of certain frequencies. By removing a rod from the lattice, we create a *cavity* that is effectively surrounded by reflecting walls. If the cavity has the proper size to support a mode in the band gap, then light cannot escape, and we can pin the mode to the defect.[1]

We illustrate this discussion of localized modes in two-dimensional photonic crystals with a system that has been studied both experimentally[2] and theoretically[3]—a square lattice of alumina columns in air. Unlike the infinitely long columns of figure 1, the real columns were sandwiched between metal plates, which introduced a cutoff TE frequency that was larger than the frequencies being investigated. The plates also insured $\mathbf{k}_{//} = 0$ propagation. In this way, the experimentalists created a system that displays only $\mathbf{k}_{//} = 0$, TM modes.

The in-plane band structure of this system is reproduced in figure 13. The system has a photonic band gap between the third and fourth bands. By examining the field patterns, as we did in the first two sections of this chapter, we would find that the first band is composed primarily of states that have no nodal planes passing through the high-ε columns. In analogy with the nomenclature of molecular orbitals, we describe

---

[1]A reader familiar with semiconductor physics can understand this result by analogy with impurities in semiconductors. In that case, atomic impurities create localized electronic states in the band gap of a semiconductor. Attractive potentials create a state at the conduction band edge, and repulsive potentials create a state at the valence band edge. In the photonic case, we can put the defect mode within the band gap with a suitable choice of $\varepsilon_{\text{defect}}$. In the electronic case, we use the effective-mass approximation to predict the frequencies and wave functions of defect modes. Within each unit cell the wave functions are oscillatory, but the oscillatory functions are modulated by an evanescent envelope. An analogous treatment is applicable to the photonic case.

[2]See McCall et al. (1991).

[3]See Meade et al. (1993a).

such a nodeless field pattern within each column as "σ-like." The second and third bands are composed of "π-like" elements, with one nodal plane passing through each column. The bottom of the fourth band has "δ-like" elements with two nodal planes per column (see insets to fig. 13). Remember that additional nodal planes in the high-ε regions correspond to larger amplitudes in the low-ε regions, which decreases the frequency.

A defect in this array introduces a localized mode, as shown in figure 14. Experimentally, this defect was created by replacing one of the columns with a column of a different radius. Computationally, the defect was introduced by varying the dielectric constant of a single column. In terms of the index of refraction $n = \sqrt{\epsilon}$, the defect varied from $\Delta n = n_{alumina} - n_{defect} = 0$ to $\Delta n = 2$ (one column completely gone). The results of the computation are shown in figure 15.

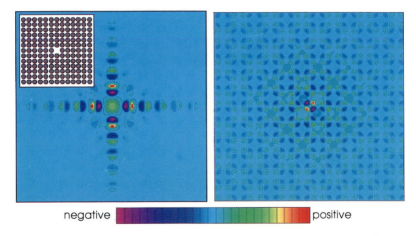

negative ▮▮▮▮▮▮▮▮▮▮▮▮▮▮▮▮▮▮▮▮▮ positive

Figure 14 The displacement fields of a state localized about a defect in a square lattice of alumina rods ($\varepsilon = 8.9$) in air. The color indicates the magnitude of the field, which is oriented in the z-direction. The defect on the left was created by *reducing* the dielectric constant of a single rod. This mode is composed of π-like elements, with one nodal line per column. Note that this defect has circular symmetry. The defect on the right was created by *increasing* the dielectric constant of a single rod. This mode is composed of δ-like elements, with two nodal lines per column. Note that this defect has δ_{xy} symmetry, i.e., it transforms like the function $f(\rho) = xy$ under rotations.

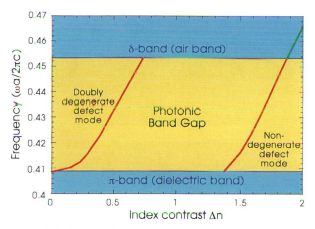

Figure 15 The evolution of a localized mode associated with a defect column in an otherwise perfect square lattice, as the defect's index of refraction decreases. An index contrast of $\Delta n = 0$ corresponds to the perfect crystal; $\Delta n = 2$ corresponds to the complete removal of one column. The horizontal lines indicate the band edges. In the gap, the frequencies (red lines) are associated with localized states, but after the line punctures the continuum it becomes a resonance with broadened frequency (green line). The state at $\Delta n = 1.58$ has the field pattern shown in the left panel of figure 14.

The photonic band gap is between the π-band of states with one nodal line, and the δ-band of states with two nodal lines in the high-ε regions. As soon as the index of refraction is less than 3, a state leaves the π-band and enters the photonic band gap. As Δn is increased between 0 and 0.8, this doubly degenerate mode sweeps across the gap. At $\Delta n = 1.4$, a nondegenerate state enters the gap, sweeps across, and penetrates the δ-band at $\Delta n = 1.8$. This mode is displayed at $\Delta n = 1.58$, before entering the δ-band, in the left panel of figure 14. Note that this state has one nodal line passing through each dielectric column, showing that it retains its π-like character as it is pushed out of the π-band. Similarly, defects with $\Delta n < 0$ pull states out of the δ band. The resulting localized states retain their δ-like character, with *two* nodal planes per column as shown on the right panel of figure 14.

The defect mode frequency increases as $\varepsilon_{\text{defect}}$ decreases, as we can see in figure 15. A simple way to understand this is to examine the effects that a small variation in $\varepsilon(\mathbf{r})$ has on the frequency of a mode. To lowest order, the change in frequency of a normalized harmonic mode **k** can be obtained using the variational theorem, equation (22) of chapter 2:

$$\delta\left(\frac{\omega}{c}\right)^2 = \int d\mathbf{r}\,\delta\left(\frac{1}{\varepsilon(\mathbf{r})}\right)|\nabla \times \mathbf{H}|^2$$

$$\frac{2\omega}{c}\delta\omega = -\int d\mathbf{r}\left|\frac{1}{\varepsilon(\mathbf{r})}\nabla \times \mathbf{H}\right|^2 \delta\varepsilon(\mathbf{r}). \tag{3}$$

For $\delta\varepsilon$ negative (i.e., removing a bit of dielectric), the corresponding frequency shift $\delta\omega$ is positive and a state can pop out of the top of the π-band (dielectric band). Increasing $\delta\varepsilon$ pushes the state deeper into the gap. Conversely, for $\delta\varepsilon$ positive, the frequency shift is negative and a state at the bottom of the δ-band (air band) can fall into the gap.[4]

Although the defect destroys the translational symmetry of the crystal, many types of defects still allow the crystal to retain some point symmetries. For instance, in the inset to figure 14, we see that after removing one column from the lattice, we can still rotate the crystal by 90° about the z-axis and leave it unchanged. If a defect does retain a point symmetry, then we can use that symmetry to classify the defect modes, just as we did in chapter 3.

For example, since the defect in the left panel of figure 14 is unchanged under 90° rotations, we can immediately predict the symmetry properties of the doubly degenerate modes that cross the gap for $0 < \Delta n < 0.8$. They must be a pair of modes that transform into each other under a 90° rotation, since that is the only doubly degenerate way to reproduce the symmetry of the surroundings.[5]

What happens to the defect mode after it reaches the δ-band? When its frequency is above the bottom of the δ-band, the defect mode is no longer trapped in the band gap; it can leak into the continuum of states that makes up the δ-band. The defect no longer creates a truly *localized*

---

[4]An analogy between this case and the case of impurities in a semiconductor might benefit the reader familiar with that subject. Since the wavelength of light is shorter in a dielectric than in air, these regions are analogous to regions of deep potential in the semiconductor. Decreasing the dielectric constant at one photonic crystal site is analogous to adding a repulsive potential to one atomic site, which pushes a state out of the π-band. Decreasing the dielectric constant further increases the repulsive character of the defect, and pushes the localized mode to higher frequencies. Conversely, increasing the dielectric region is analogous to adding an *attractive* potential that pulls a state out of the δ-band.

[5]In the language of group representation theory: the defect has the symmetry of the C_{4v} point group, and the only doubly degenerate representation of that group is a pair of π-like functions.

mode. However, since we expect a smooth transition between localized and continuum behavior, the mode still concentrates much of its field energy near the defect. But in this case, the defect is not surrounded by reflecting walls, so the energy will leak away into the continuum of states at some rate Γ. We call such a mode a *resonance*.[6] The peak that the defect creates in the crystal's density of states widens in proportion to Γ; as the mode penetrates farther into the continuum, farther and farther from the true bound state, the resonance broadens away and melts into the continuum.

Planar Localization: Surface States

The majority of our discussion has concerned the interior of photonic crystals of (presumably) infinite extent. But real crystals are necessarily bounded—so what happens at the surface of a two-dimensional photonic crystal? In this section, we explore the *surface modes* that photonic crystals are capable of sustaining. In a surface mode, light is localized at a surface plane, as shown schematically in figure 12. The field amplitudes decay exponentially away from the surface.

We can characterize a given surface by its *inclination* and its *termination.* Surface inclination specifies the angles between the surface normal and the crystal axes. Surface termination specifies exactly where the surface cuts across the unit cell; for example, we can end a two-dimensional lattice of circles by stopping after some whole number of circles, or by cutting each circle in half at the boundary, or by cutting off some arbitrary fraction.

We will focus on the surface states of a square lattice of dielectric columns. Many of the arguments and results that we present, however, are quite general. Specifically, we will return to the square array of alumina rods, with $\varepsilon = 8.9$. Consider the TM band structure, which has a photonic band gap between the first and second bands. For our surface inclination, we choose planes of constant x.[7] We will look at

---

[6]This name is chosen to suggest the correspondence with the phenomenon of resonance scattering in optics and quantum mechanics. In a scattering experiment, tuning the incident energy near the energy of a true bound state causes a peak in the cross section due to the resonance of the incident beam with the potential. Here, there is a peak in the density of states because our defect mode is very close to a true localized mode. The rate of decay of the resonance is proportional to the peak width in both cases.

[7]This is known as the (10) surface of the square lattice.

two different terminations—either we will draw the boundary just outside a line of whole columns, or we will cut the outermost columns in half. These two terminations are depicted in the insets to figures 16 and 17, respectively.

Because the translational symmetry in the x-direction is ruined, we can no longer describe the electromagnetic modes by a wave vector k_x. However, the system still has discrete translational symmetry in the y-direction, and continuous translational symmetry in the z-direction. We can still classify the modes of the surface Brillouin zone with wave vectors k_y and k_z, which are bounded by $-\pi/a < k_y \leq \pi/a$ and $-\infty < k_z < \infty$.

Modes propagating in the z-direction become guided, and the corresponding bands have a decaying bandwidth, as discussed in the section entitled "Out-of-Plane Propagation." In this case, surface modes become guided modes, as in the case of the multilayer film. These two cases are closely related because the systems are homogeneous in the direction of propagation. For this reason, we limit the following discussion to in-plane propagation, for which $k_z = 0$.

As in the case of defects, *surface modes* occur when there are electromagnetic modes near the surface, but they are not permitted to extend into the crystal at that frequency because of a photonic band gap. But with surface modes we must consider the modes' behavior not just as a function of frequency, but also as a function of k_y. A surface mode must have the appropriate combination (ω_0, k_y) that is disallowed in the crystal, not just the appropriate ω_0.

To determine where these regions exist, we pick a specific (ω_0, k_y) and ask if there is any k_x which will put that mode on a band. That is, by a suitable choice of k_x, can we arrange for some band n that $\omega_0 = \omega_n(k_x, k_y)$? If we can, then there is at least one extended state in the crystal with that combination (ω_0, k_y). If we tried to set up a surface mode with those parameters, it would not be localized; it would leak into the crystal.

This process of searching all possible k_x for each k_y is called "projecting the band structure of the infinite crystal into the surface Brillouin zone." We take all of the information from the full crystal band structure, and extract the information relevant for the surface. A true, localized, surface mode must be evanescent both inside the crystal and outside, in the air region, so we must project the band structure of both the photonic crystal and the air region.

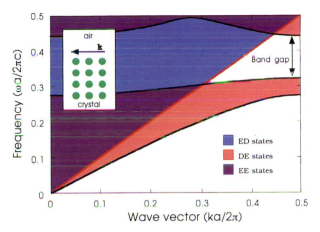

Figure 16 The projected band structure of the constant-*x* surface of the square lattice of alumina rods in air. The shading denotes regions in which light is transmitted (purple), internally reflected (red) and externally reflected (blue). The crystal is terminated as shown in the inset.

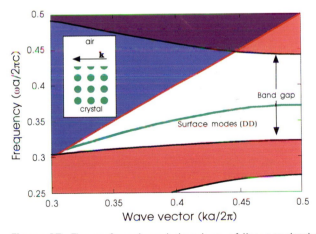

Figure 17 The surface band structure of the constant-*x* face of the square lattice of alumina rods in air. The shading denotes regions in which light is transmitted (purple), internally reflected (red) and externally reflected (blue). The line in the gap corresponds to a surface bands in which light is exponentially localized to the surface (green). The crystal is terminated as shown in the inset.

Figure 16 shows the projected band structure of the constant-x surface of the square lattice of dielectric rods. In order to understand it, we first consider the projected band structures of the outside air and photonic crystals separately. As before, we label each section of the plot with two letters; the first tells whether the states are Extended or Decaying in the air region, the second letter tells the same for the crystal region. The union of regions *EE* and *ED* shown in figure 16 is the projection of the free light modes onto the surface Brillouin zone. For a given k_y, there are light modes at all frequencies $\omega \geq c|k_y|$. Along the line $\omega = ck_y$, the light travels parallel to the surface, and increasing ω corresponds to increasing k_x. Similarly, the union of regions *EE* and *DE* represents the projected band structure of the photonic crystal. Note that the photonic crystal contains a band gap at $0.32 < (\omega a/2\pi c) < 0.44$.

Now we can understand the three types of surface states of the projected surface Brillouin zone: light that is transmitted (*EE*), light that is internally reflected (*DE*), and light that is externally reflected (*ED*). In the region of (ω, k_y) marked *EE,* the modes are extended in both the air and in the crystal, so it is possible to transmit light with those parameters through the crystal. In the *DE* region, there are modes in the crystal, but they are beneath the band edge of the air states. Thus the light can extend into the crystal, but exponentially decays into the surrounding air. This is nothing but the familiar phenomenon of total internal reflection. In the *ED* region, the situation is reversed. There, the modes can extend into the air, but decay away into the crystal.

Finally, there might exist bona fide surface modes, which decay away from *both* sides of the surface (labeled *DD*). Such a mode is displayed in figure 17, which shows the band structure of the constant-x surface terminated by cutting the columns in half. The modes in the *DD* region are below the band edge of the air modes, and are also within the band gap of the crystal. The fields decay exponentially in both directions, which pins it to the surface plane. By exciting those modes, we can imprison light at the surface of the crystal.

Further Reading

Appendix A catalogs many of the analogies between the field of photonic crystals and the disciplines of quantum mechanics and solid-state physics. Appendix B provides a more detailed discussion of the Brillouin zones of the crystal geometries we have studied in this chapter. Appendix C provides the band gap locations for a wide variety of two-dimensional photonic crystals.

Further information about calculating the effective dielectric constants of a medium is provided in Aspnes (1982). Surface states on the interface between two different materials are reported in Meade et al. (1991b). The experimental investigation of the square lattice of dielectric columns for bulk states can be found in Robertson et al. (1992) and for surface states in Robertson et al. (1993). Some early experimental and theoretical approaches to two-dimensional systems are in McCall et al. (1991), Smith et al. (1993), Plihal and Maradudin (1991), and Villeneuve and Piche (1992).

Meade et al. (1991a,b) contains a systematic treatment of the square lattices of dielectric rods and veins. Winn et al. (1993) contains a more systematic treatment of square and triangular lattices of columns.

6

Three-Dimensional
Photonic Crystals

The optical analog of an ordinary crystal is a three-dimensional photonic crystal—a dielectric that is periodic along three different axes. Three-dimensional photonic crystals can have the novel properties we discussed in the previous two chapters, including band gaps, defect modes, and surface states. In this chapter we will present two examples of three-dimensional crystals with complete band gaps: a diamond lattice of air holes, and a drilled dielectric known as Yablonovite. Inserting defects still allows us to localize light in a plane or on a line, but in three dimensions we have the additional freedom to create *guided* linear modes and modes that are localized at a single point.

Two Classes of Photonic Crystals

There are an infinite number of possible geometries for a three-dimensional photonic crystal, but we are particularly interested in those geometries that promote the existence of photonic band gaps. The results of the previous chapter give us a hint to try structures that contain connected networks of dielectric spots. In three dimensions, we might try creating our crystals with dielectric tubes and spheres, analogous to the spots and veins of our successful two-dimensional crystals. We will investigate two possibilities.

The first type is created by taking a three-dimensional lattice and placing a sphere at each lattice point. We can completely characterize crystals of this type by the lattice vectors, the dielectric constants of the spheres and the embedding material, and the radius of the spheres. The upper portion of figure 1 shows an example of this type of crystal, with dielectric spheres arranged in a diamond lattice. We can also

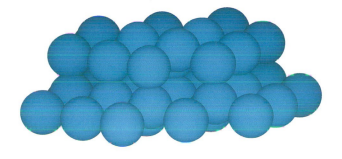

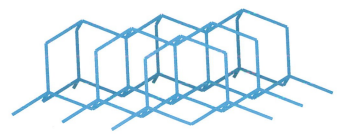

Figure 1 Two examples of three-dimensional photonic crystals. The upper figure shows a crystal consisting of dielectric spheres at the sites of a diamond lattice. The lower figure shows a crystal formed by connecting the sites of a diamond lattice by dielectric rods.

reverse the dielectric constants, and place air bubbles in a dielectric material.

The second type results from taking a lattice and connecting the lattice points with cylindrical columns. The lower portion of figure 1 shows a crystal of this type; it is a diamond lattice whose lattice points have been joined to form a network of tubes. Structures of this type have been realized experimentally by drilling a regular pattern of holes into a solid block of dielectric. This type of crystal can be characterized by the dielectric constants of the different regions, the pattern and angles of the drilling, and the radius of the holes.

In both of these cases, the crystals are composed of only two different dielectric constants. We argued in chapter 2 that scaling the entire dielectric function by some constant factor, $\varepsilon(\mathbf{r}) \rightarrow \varepsilon(\mathbf{r})/s^2$, results in a trivial rescaling of the band structure, $\omega \rightarrow s\omega$. Therefore, what is really important is the ratio of the two different dielectric constants, not the actual values. As long as we keep the ratio fixed, we can scale the dielectric constants to whatever level we like, and the optical properties will be essentially the same. We define the *dielectric contrast* as the ratio of the dielectric constants of the high-ε and low-ε regions: $\varepsilon_{high}/\varepsilon_{low}$.

Generally speaking, band gaps tend to appear in structures with a high dielectric contrast. The more significant the scattering of light, the more likely a gap will open up. One might wonder whether *any* geometry could have a photonic band gap, with a sufficiently high dielectric contrast. This is in fact the case for most two-dimensional crystals.

For three-dimensional crystals, complete photonic band gaps are rarer. The gap must smother the entire three-dimensional Brillouin zone, not just any one plane or line. For example, in figure 2 we show the band structure for a face-centered cubic lattice of air spheres in a high dielectric ($\varepsilon = 13$) medium. Although the dielectric contrast is very large, there is no complete photonic band gap. Interestingly, there is a large space between bands 2 and 3 throughout most of the Brillouin zone, but the distribution of mode frequencies around the U and W wave vectors prevent the gap from being complete.

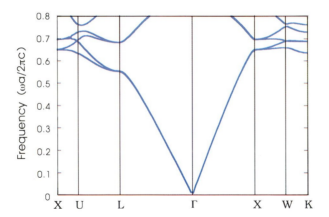

Figure 2 The photonic band structure for the lowest-frequency electromagnetic modes of a face-centered cubic lattice of air spheres in a dielectric ($\varepsilon = 13$) medium. Note the absence of a complete photonic band gap. The wave vector varies across the irreducible Brillouin zone, from Γ to X to W to K, then back to Γ through X, U, and L. See appendix B for a discussion of the Brillouin zone for a face-centered cubic lattice.

Nevertheless, several three-dimensional crystals have been discovered that do yield sizable complete photonic band gaps. These particular crystals will be taken up in the next section. In most of the studies undertaken to date, as the dielectric contrast increases for a given crystal, a photonic band gap opens up only after some nonzero threshold is reached, and then monotonically increases in width.

Crystals with Complete Band Gaps

Ho, Chan, and Soukoulis were the first theorists to correctly predict that a particular three-dimensional photonic crystal would have a complete band gap.[1] Their crystal was of the first type of crystals discussed earlier—a diamond lattice of spheres, similar in form to the one shown in the upper panel of figure 1. They found that a complete photonic band gap exists whether one embeds dielectric spheres in air or air spheres in a dielectric medium, as long as the sphere radius is chosen appropriately.

The band structure of the lattice of air spheres is shown in figure 3. To maximize the size of the band gap, the sphere radius r is chosen to be $0.325a$, where a is the lattice constant. Between the second and third bands resides a band gap with a gap-midgap ratio of 0.29.

Most of this structure (81% by volume) is air. In fact, the diameter of the air spheres is larger than the distance between spheres, $0.65a > \sqrt{3}/2a$, so the air spheres overlap. Both the air and the dielectric regions are connected, in the sense that there are no isolated spots of either. We can think of this crystal as two interpenetrating diamond lattices, one composed of connecting air spheres and one composed of connecting dielectric remnants.

A different structure, which has proven simpler to manufacture in the laboratory than the true diamond structure, consists of a dielectric medium that has been drilled along three of the axes of the diamond lattice, as shown in figure 4. It has been named *Yablonovite*, after its discoverer E. Yablonovitch. Yablonovite has been built on the microwave lengthscale and has the distinction of being the first three-dimensional photonic crystal with a complete photonic band gap to be fabricated.[2]

Drilling holes with radius $r = 0.234a$ results in a structure whose

---

[1]See Ho et al. (1990) for the original paper.
[2]This is reported in Yablonovitch et al. (1991a).

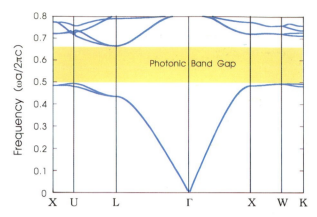

Figure 3 The photonic band structure for the lowest six bands of a diamond lattice of air spheres in a high dielectric (ε = 13) material. The wave vector varies across the irreducible Brillouin zone, from Γ to X to W to K, then back to Γ through X, U, and L. See appendix B for a discussion of the Brillouin zone for a face-centered cubic lattice.

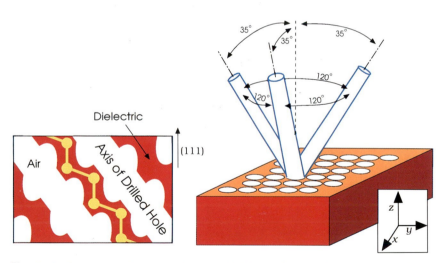

Figure 4 The method for constructing Yablonovite: a slab of dielectric is covered by a mask consisting of a triangular array of holes. Each hole is drilled three times, at an angle of 35.26° away from the normal, and spread out 120° on the azimuth. This results in a three dimensional structure whose ($1\overline{1}0$) cross section is shown on the left. The dielectric connects the sites of a diamond lattice, shown schematically in yellow. The dielectric veins oriented vertically (111) have greater width than those oriented diagonally ($11\overline{1}$).

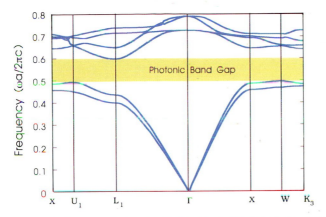

Figure 5 The photonic band structure for the six lowest bands of Yablonovite. A detailed discussion of this band structure can be found in Yablonovitch et al. (1991a).

photonic band gap has a gap-midgap ratio of 0.19, as shown in figure 5. Like the diamond lattice of air spheres, we can think of Yablonovite as two interpenetrating diamond lattices, one a connected region of dielectric and the other a connected region of air.[3]

Localization at a Point Defect

Now that we have introduced two structures that have photonic band gaps, we can discuss some of the novel features that result. We have already seen that defects in photonic crystals can localize light modes. In one dimension, this meant we could confine light to a single defect plane, and in two dimensions we could localize light at a linear defect. In three dimensions, we can perturb a single lattice site, and thereby trap light at a single point in the crystal. With the point defect, we pull a state from the continuum above or below the gap *into* the gap itself, and a localized mode results.

[3] A note for experts: Yablonovite is not a *true* diamond structure, which would require drilling holes down the six axes (110), (101), (011), ($\bar{1}$10), ($\bar{1}$01) and (0$\bar{1}$1). By drilling along only three axes, the (111) direction is singled out. For this reason, Yablonovite does not have the full diamond symmetry, but only a D_{3d} symmetry (a threefold rotation axis (111) as well as mirror planes and inversion symmetry). The lowering of symmetry is responsible for broken degeneracies at some of the special points of the Brillouin zone. Nevertheless, Yablonovite is "diamondlike"—it has the same topology as a diamond lattice. It consists of a set of dielectric veins that connect the sites of a diamond lattice. However, because only three rods are drilled, the veins along the (111) direction have a larger diameter than those along the ($\bar{1}\bar{1}$1), (1$\bar{1}\bar{1}$), and ($\bar{1}$1$\bar{1}$) directions.

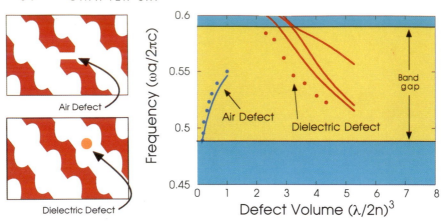

Figure 6 Plotted are the frequencies of the localized modes of Yablonovite as the defect size varies. The dots indicate measured values (Yablonovitch et al. 1991b), the lines indicate computed values (Meade et al. 1993a), and the yellow region is the photonic band gap. The modes on the blue line result from an air defect, while the modes on the red lines result from a dielectric defect. The defect volume is expressed in units of $(\lambda/2n)^3$, where λ is the midgap vacuum wavelength and n is the index of refraction of the dielectric material.

Two simple ways to perturb a single lattice site are to add extra dielectric material where it does not belong, or to remove some of the dielectric material that should be there. We might call the first a "dielectric defect," and the second an "air defect." Examples of both are illustrated in figure 6. These defects are similar to the ones we explored in chapters 4 and 5, and our discussion will simply develop and extend those earlier results.

By inserting a point defect, we ruin the discrete translational symmetry of the lattice, so (strictly speaking) we can no longer classify the modes of the system with a wave vector **k**. Instead, we focus on the crystal's density of states. The defect causes the appearance of a single peak of newly allowed states into the density of states at a frequency that may lie in the photonic band gap. The width of this peak tends to zero as the crystal size tends to infinity. Since no extended states are allowed in the crystal within the band gap, the new peak must consist of localized states. Simply put, modes in the band gap decay exponentially away from the defect. But in this case they decay exponentially in all *three* dimensions—they are trapped near a single point.

Why does a defect localize electromagnetic modes? We remind the reader of the simple, intuitive picture: the defect is like a cavity with perfectly reflecting walls. If light with a frequency within the band gap somehow winds up near the defect, it cannot leave, because the crystal does not allow extended states at that frequency. Therefore if the defect allows a mode to be excited with a frequency within the band gap, that mode is forever trapped.

These localized modes can be realized experimentally. We illustrate our discussion by reporting the results of inserting defects into Yablonovite, a system that has been studied both computationally and experimentally. The left panel of figure 6 depicts the Yablonovite defects (both air and dielectric) that were studied. By systematically measuring microwave transmission levels, the experimenters mapped the frequencies of the defects, which are shown in the right-hand panel of figure 6. The theoretical values are also shown.

The defects create localized modes within the photonic band gap. An air defect introduces a single, nondegenerate state into the photonic band gap, which crosses from the dielectric to the air band as the defect frequency is raised. The field patterns of this state are shown in figure 7. Light is localized about the defect in a region like an inner tube—a toroidal geometry. The magnetic field lines "flow" around the inside of the torus, while the displacement field circulates around the magnetic field, on the surface of the torus. The dielectric defect, on the other hand, introduces states that cross in the opposite way—from the air band to the dielectric band.

Figure 6 shows that the defect frequency is an increasing function of the volume of the air defect, and a decreasing function of the volume of the dielectric defect. We found a similar result in the two-dimensional case, and the reason is the same (see chapter 5).

There is one significant difference in the three-dimensional case, however. The defect must be larger than some critical size to localize light; as the defect is increased from zero size, it must pass through some nonzero threshold before its localizing power begins. In one- and two-dimensional crystals, we found that even arbitrarily small defects can localize modes.[4]

---

[4]This is the electromagnetic analog of a famous theorem of quantum mechanics. The theorem states that an arbitrarily weak attractive potential can bind a state in one and two dimensions, but not in three dimensions.

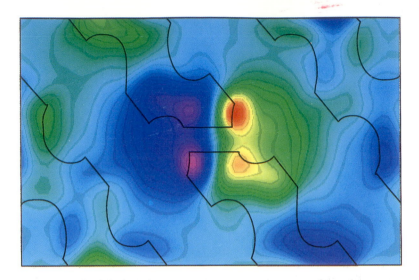

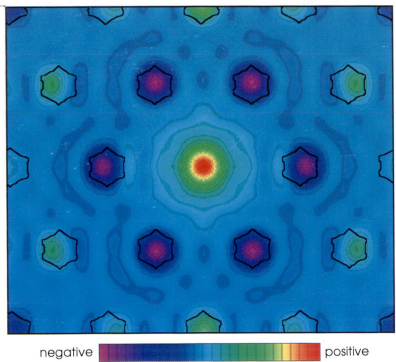

negative ▮▮▮▮▮▮▮▮▮▮▮▮▮▮▮ positive

Figure 7 Two views of a state localized near an air defect in Yablonovite. The upper figure shows a slice through the ($1\bar{1}0$) plane, the same cross-section shown inset in figures 3 and 6. The dielectric, designated by solid lines, has the same outline as in figure 6. Note the air defect at the center. The color indicates the strength of the magnetic field, which is oriented out of the page along the ($1\bar{1}0$) direction. The lower figure shows a slice through the (111) plane passing through the defect's center. This plane cuts through the vertical veins of Yablonovite. Note the air defect at the center. The color indicates the displacement field, which is primarily oriented out of the page, along the (111) direction.

Although a point defect destroys the translational symmetry of the lattice, many defects retain point symmetries about their centers. For instance, in the lower panel of figure 7 we see that even after introducing the air defect into the crystal, we retain the threefold rotation symmetry of the crystal. We could still classify the modes, including the new localized modes, by how they transform under a 120° rotation in the plane of the picture. Other symmetries, like mirror reflections and inversions, may also be left intact by a particular defect.

Localization at a Linear Defect

Another class of defects are *linear* defects, which extend in one direction. Two typical examples for the Yablonovite crystal are shown in figure 8. On the left is an air defect, in which a column of dielectric material has been removed from the crystal. On the right is a dielectric defect, in which a dielectric column has been added to the air region.

One might think of these as linear arrays of point defects. By choosing a proper radius and orientation for a line defect, it is possible to create a defect *band* with frequencies in the photonic band gap of the crystal. The states in this band extend along the defect, but decay exponentially into the rest of the crystal.

By aligning the line defect with one of the translation vectors of the crystal (as in fig. 8), we preserve the translational symmetry along this one direction. Because of this, we can classify the defect modes with a defect wave vector k, which characterizes the phase variation along the defect line. Such states transport electromagnetic energy along the line defect. This fact suggests that linear defects are analogous to

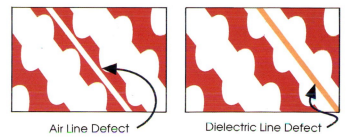

Air Line Defect   Dielectric Line Defect

Figure 8 Schematic illustration of an air line defect (left) and a dielectric line defect (right) in Yablonovite.

metallic waveguides. Light is trapped in a tube with dimensions comparable to its wavelength, and with perfectly reflecting walls. An example of such a waveguide is presented at the end of chapter 7.

Localization at the Surface

As before, we will discuss the effect of terminating a photonic crystal by examining its surface band structure. Again we will search for electromagnetic surface modes, which decay away exponentially on both sides of the surface plane. In fact, although we focus on surfaces, these types of localized states may occur at planar defects within the crystal as well.

Suppose we terminate a 3-D photonic crystal in the z-direction. By doing this, we destroy the translational symmetry in that direction, so we can no longer classify the states of the crystal with a definite k_z. The crystal still has translational symmetry parallel to the surface, so the electromagnetic modes *do* have a definite $\mathbf{k}_{//}$. We must project the full three-dimensional band structure onto the surface Brillouin zone, with the procedure we described in the last section of chapter 5.

We illustrate our discussion of the general features of surface states with the (111) surface of Yablonovite,[5] depicted in cross section in figure 9. As discussed earlier, the dielectric veins oriented vertically have greater width than those oriented diagonally. Note that this system is invariant under 120° rotations about an axis through one of the vertical veins.

Before we consider the band structure of the interface, let us first consider the projected band structures of the air and photonic crystals separately. As before, we use an "E" to stand for regions of extended states, and a "D" for regions of decaying states; states are labeled by either E or D on both the air and the crystal sides.

For example, the union of regions EE and ED shown in figure 9 is the projection of the extended air states ($E\_$) onto the surface Brillouin zone of the crystal ($\_E$ or $\_D$). For a given $\mathbf{k}_{//} = (k_x, k_y)$, there are extended light modes for all frequencies $\omega \geqslant ck$. Along the line $\omega = ck$, the light travels parallel to the surface, and increasing ω corresponds to increasing k_z. Likewise, the union of regions EE and

---

[5]We will use Miller indices to refer to crystal planes. Readers unfamiliar with this notation will find a brief description in appendix B.

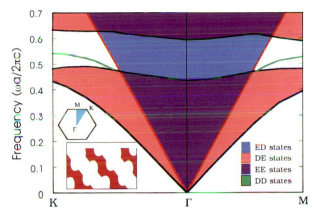

Figure 9 The band structure of the (111) surface of the Yablonovite crystal along special directions in the surface Brillouin zone. The shading denotes regions in which light is transmitted (purple), internally reflected (red) and externally reflected (blue). The green lines in the gap correspond to surface bands in which light is localized at the surface. The surface Brillouin zone is shown at left, with the irreducible zone shaded light blue. The M point is a distance $\sqrt{2/3}\,(2\pi/a)$ away from Γ along $(1,\bar{2},1)$. The K point is a distance $(\sqrt{8}/3)\,(2\pi/a)$ away from Γ along $(0,\bar{1},1)$. This surface band structure corresponds to a termination of $\tau = 0.75$. A $(1\bar{1}0)$ cross-section is inset, with air above and photonic crystal below. Dielectric regions are shaded brown.

DE represents the projected band structure of the photonic crystal. Note that the photonic crystal contains a gap at $0.49 < \omega a/(2\pi c) < 0.59$, in which no extended states are allowed.

We catalog the surface states as one of four types: transmitted (*EE*), internally reflected (*DE*), externally reflected (*ED*), and bona fide surface modes (*DD*). In the region of ($\mathbf{k}_{//}$, ω) marked *EE*, the modes are extended in both the air and in the dielectric, so it is possible for light to traverse the crystal. In the *DE* region, modes extend in the crystal, but they are beneath the band edge for air states. In the *ED* region, the situation is reversed; there are extended states in the air but they lie in the gap of the crystal. Finally, in the region marked *DD*, the states are below the band edge of the light in the air, as well as in the gap of the crystal. The light decays exponentially in both directions away from the surface plane.

The fields associated with the zone edge surface mode (at the

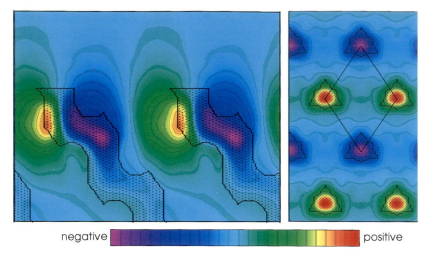

negative | positive

Figure 10 Two views of the fields of the surface mode at the Brillouin zone edge (M) for a surface termination of τ = 0.75. The dielectric regions are shaded. *Left:* A view in the (1 1̄ 0) plane. The color indicates the magnitude of the magnetic field **H**. The displacement field **D** points along the lines of constant **H**, and is large where **H** changes rapidly. In this cross-section, **k** lies in the plane of the page. *Right:* A view of the (111) plane passing through the top layer of the surface. **D** is primarily in the plane. The color indicates the strength of **H**, which is primarily normal to the plane. The surface unit cell is marked with a parallelogram. Fields of neighboring unit cells have opposite signs, as expected for a state with a wave vector at M. In this view, **k** is directed up.

M-point of the Brillouin zone) are shown in figure 10. The panel on the left is a view in a (110) cross section through the surface plane. Because the crystal also contains a mirror symmetry through this plane, we expect the fields to *appear* as TE or TM. In this case, **D(r)** lies in the plane and **H(r)** is everywhere normal to the plane. The fields are strongly localized in the plane of the surface, barely extending past the uppermost dielectric layer. As usual, most of the power of the field is concentrated in the high-ε regions.

The panel on the right of figure 10 displays the fields associated with the surface mode in a (111) plane passing through the top layer of the surface. Although this is not a mirror plane, **H(r)** is primarily in the plane, as shown by the vectors, and **D(r)** is primarily normal to the plane. For this reason we may describe the mode as "TE-like." Of

course, none of the surface modes of this photonic crystal are rigor-
ously TE nor TM.[6]

Until now, we have considered the band structure of the particular
choice of surface termination shown in the inset to figure 9. However,
the surface can be terminated in a variety of ways, as shown in the
insets to figure 11. We must specify not only the inclination of the
surface plane, but also where in the unit cell we will cut the crystal.

We introduce a termination parameter τ that varies between 0 and 1
to describe the termination of any surface: let $\tau = 0$ when the surface
is terminated through the vein center, as shown in the inset to the
bottom panel of figure 11, and let τ increase linearly as the height of
the surface is raised. By the time $\tau = 1$, we have cut across the next
bond center, so the same termination as $\tau = 0$ results.

The surface band structure varies in an interesting manner as the
surface termination changes, as is shown in figure 11. As the termination parameter τ increases, more dielectric is added, and the frequency
of the surface mode lowers accordingly. Bands sweep down from the
air band to the dielectric band, which may be either TE-like or TM-like.

Consider the band structure as τ is increased from $\tau = 0$ to $\tau = 1$.
As the termination is increased, exactly two states are swept from the
air band to the dielectric band—one is TE-like, and the other is
TM-like. By increasing the termination of the surface from $\tau = 0$ to
$\tau = 1$ we have added one bulk unit cell per surface unit cell, and have
increased the total number of states in the dielectric band by two. This
is true for each point in the Brillouin zone.

This suggests a general claim: *for a crystal with a band gap, and a
surface of a given inclination, we can always find some termination
that allows localized surface modes.* We sketch the argument briefly:
since the crystal as a whole has a gap, the surface Brillouin zone must

---

[6]To understand why, consider the simpler case of a surface wave atop a multilayer film, as
presented in the last section of chapter 4. In this case the TE/TM classification is exact, as we
can see from a symmetry argument. The surface of a multilayer film has continuous translational
symmetry parallel to the surface plane, so we can label the states by in-plane wave vector **k**.
Note that the plane which contains both **k** and the surface normal is a mirror plane. As discussed
toward the end of chapter 3, fields must be either even or odd with respect to this mirror symmetry
to *any* choice of origin—therein lies the TE/TM distinction. But at this particular surface of
Yablonovite, the symmetry is lower, and the mirror plane only exists for *specific* choices of origin.
We can no longer argue that modes must be TE or TM although they will *appear* to have this
symmetry when viewed in the mirror plane itself.

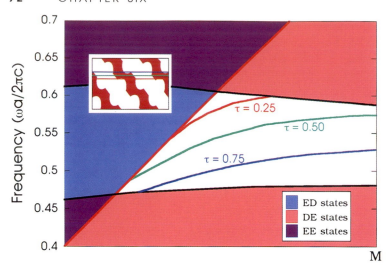

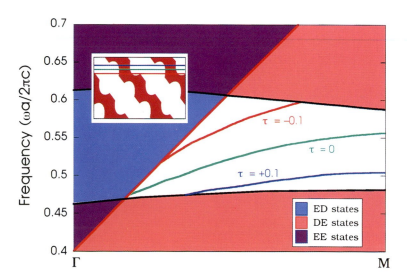

Figure 11 The surface band structures of TE-like (top) and TM-like (bottom) modes. This figure zooms in on the gap region of figure 9, using the same classification scheme *EE, ED, DE, DD*. The insets indicate the corresponding surface terminations, shown with colored lines. Larger values of **τ** are for terminations higher up in the unit cell and therefore more dielectric material near the surface. Correspondingly, a higher **τ** yields a lower surface band frequency.

also have a gap. As before, we introduce a termination parameter τ so that as τ varies from 0 to 1, there are b crystal unit cells introduced per surface unit cell. There must then be $2b$ new states transferred from the air band to dielectric band. As the frequencies of these states decrease from the bottom of the air band to the top of the dielectric band, they sweep through the gap—these are localized surface states. This argument for the existence of surface states also applies to the case of multilayer films that lack a gap covering the full three-dimensional Brillouin zone, but do have a gap in the direction normal to the surface.

Further Reading

Appendix B describes the reciprocal lattice and the Brillouin zone in more detail, including the reciprocal lattice of the face-centered cubic lattice. Occasionally in this chapter we use Miller indices to refer to crystal planes and axes. Interested readers can find a brief description of this notation in appendix B. For a more complete treatment of these topics, consult the first few chapters of a text on solid-state physics (for example, Kittel 1986).

The first suggestion that a material with a complete photonic band gap in three dimensions might be fabricated appeared in Yablonovitch (1987). The history of the discovery of the first material with a complete photonic band gap can be followed in Yablonovitch and Gmitter (1989), Satpathy et al. (1990), Leung and Lin (1990), Zang and Satpathy (1990), Ho et al. (1990), Chan et al. (1991), and Yablonovitch et al. (1991a).

Yablonovitch (1987) has also suggested that a crystal with a complete three-dimensional photonic band gap would totally inhibit the spontaneous emission of atoms inside the crystal, for the frequency range of the gap. This has been known to occur in microwave cavities (Kleppner 1981). A partial inhibition has been observed at visible frequencies by Martorell and Lawandy (1990).

7

Designing Photonic Crystals for Applications

In the first few chapters, we assembled a number of theoretical tools to help understand the properties of photonic crystals. The preceding three chapters focused on the central question of the field: *Which structures have interesting properties, and why?* This chapter will be concerned with a different question altogether: What can we *do* with photonic crystals, now that we understand them?

Our introduction promised that a technological revolution would accompany our ability to control the propagation of light. In this chapter we will sketch out designs for a few basic components of light-controlling devices that employ photonic crystals. For simplicity, we will work with two-dimensional systems, although the ideas generalize easily to the cases of one and three dimensions.

Instead of tying ourselves to a particular application, we will use generic examples that are sure to find use in many experiments and devices: a perfect *dielectric mirror*, which reflects light; a *resonant cavity*, which traps light; and a *waveguide*, which transports light. Not only do these examples demonstrate the promise of photonic-crystal technology, but they also provide a perfect setting to review the concepts we introduced in previous chapters.

A Reflecting Dielectric

Long ago, engineers solved the problem of controlling light propagation in the microwave regime[1] by using metallic components to guide, reflect, and trap light. These components rely on the high

---

[1]The "microwave regime" includes light with wavelengths in the range from about one millimeter to about ten centimeters.

conductivity of metals, a rather complicated electronic property that may depend strongly on frequency. Unfortunately, for light of higher frequency (like visible light, for instance), metallic components suffer from high dissipative losses.

As we have seen, the reflectivity of photonic crystals derives from their geometry and periodicity, not a complicated atomic-scale property. The only demand we make on our materials is that for the frequency range of interest (which is often a narrow band), they should be essentially lossless. Such materials are widely available all the way from the ultraviolet regime to the microwave. We saw in chapter 2 that the photonic properties scale easily with frequency and ε, so devices made at one scale are sure to work at other scales.

Since the heart of so many devices is reflectivity, our first task will be to design a two-dimensional crystal that reflects all in-plane light within some specified frequency band, without appreciable absorption. Once finished, we could use this crystal in a band-stop filter. Or, since the band structures of two-dimensional photonic crystals are different for TE and TM light, we could employ it as a polarizer. Or, as we will see in the next two sections, we can use it to make resonant cavities and dielectric waveguides.

For concreteness, we will design elements for a particular wavelength of light: $\lambda = 1.5$ μm, the wavelength of light which is often used in telecommunications.[2] Suppose we construct the crystal from gallium arsenide (GaAs), a material widely used in optoelectronics. For light with a wavelength between $\lambda = 1.0$ μm and $\lambda = 10.0$ μm, GaAs has a dielectric constant of 11.4.[3] Can we design a photonic crystal to meet these specifications?

In order to make a reflecting structure, we need to choose a crystal geometry that provides a photonic band gap. We should also choose a geometry that is relatively easy to fabricate at micron-level dimensions. After consulting an atlas of gap maps, such as the abbreviated one provided in appendix C, we notice a particularly simple geometry with those characteristics: the triangular lattice of air columns. It has band gaps for both TE and TM modes, it has an overlapping band gap for both polarizations, and we can make it by simply etching holes into a GaAs sheet. The gap map is reproduced in figure 1.

---

[2]For an introduction to optoelectronics see Yariv (1985).
[3]As reported in Pankove (1971).

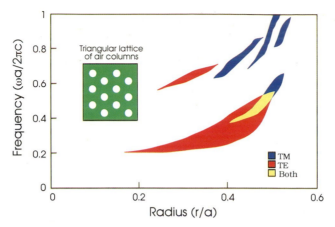

Figure 1 The gap map for a triangular lattice of air columns in GaAs. Plotted are the locations of the photonic band gaps as the radius of the columns increases. Notice the overlapping TM (blue) and TE (red) band gaps around $r/a = 0.45$ and $\omega a/2\pi c = a/\lambda = 0.5$.

If we wanted to make a polarization-sensitive device, we would choose a band gap for either TE or TM modes alone. Here, we will choose the overlapping region on the gap map, in which all in-plane light is reflected, regardless of polarization. That region's thickest extent (largest band gap) occurs when $r/a = 0.45$, and the gap is centered around $\omega a/2\pi c = 0.5$. We have chosen $\lambda = 2\pi c/\omega = 1.5$ μm, so we need

$$\frac{\omega a}{2\pi c} = \frac{a}{\lambda} = \frac{a}{(1.5\mu m)} = 0.5 \qquad a = 0.75\mu m \, . \tag{1}$$

The condition that the wavelength is 1.5 μm has provided a, and now we use the condition that there is a complete band gap to determine r. Setting $r/a = 0.45$, we find $r = 0.34$ μm. Our structure is now determined completely: we will use a GaAs substrate in which air holes of radius 0.34 μm have been etched in a triangular pattern, with lattice constant 0.75 μm.

As we can see from figure 1, the extent of the band gap is from $\omega a/2\pi c = a/\lambda = 0.45$ to $\omega a/2\pi c = 0.55$. The gap-midgap ratio is $0.1/0.5 = 20\%$. The frequency band corresponds to a wavelength band from $\lambda = 0.73$ μm to $\lambda = 1.7$ μm. This region adequately covers the frequency range of interest.

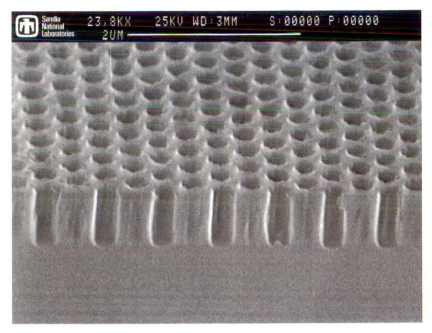

Figure 2 A scanning electron microscope image of a triangular array of air columns in gallium arsenide. The columns have a radius of 122.5 nm, the lattice constant is 295 nm, and the columns are about 600 nm tall. This structure was fabricated by Wendt and co-workers (1993).

This particular design has actually been fabricated in a laboratory.[4] Figure 2 shows a micrograph of the triangular lattice of air columns. The etching was performed with electron-beam lithography. Experimental tests have verified a variety of the optical properties that we would expect from our considerations in this text.

A Resonant Cavity

Now that we have designed a perfectly reflecting dielectric, what else can we do with it? In previous chapters, we have studied the phenomena that take place in the vicinity of a defect in a photonic crystal. We can put these discussions to good use by designing a resonant cavity in a photonic crystal.

---

[4]See Wendt et al. (1993) for the report of this achievement.

As we have seen, defects admit the existence of localized modes within a very narrow frequency band. For example, if we place a defect in our triangular lattice of air columns, and we excite a mode with a frequency within the band gap (at $\lambda = 1.5$ μm, for instance), the light will have nowhere to go. It will be trapped by perfectly reflecting walls. Of course, our structure will only confine light in the plane of periodicity. To prevent it from escaping in the third direction, another method would be needed—one might sandwich the triangular lattice between two metallic plates, or two dielectric slabs (counting on total internal reflection). Alternatively, one could use a fully three-dimensional photonic crystal. But for pedagogical purposes, we will remain focused on the two-dimensional case.

Why would we want to create a defect mode? One obvious answer is to serve as a resonant cavity. Such cavities are crucial components of laser systems. A defect mode in a photonic crystal would serve as an effective resonant cavity, since it would only trap light in a very narrow frequency band and would hardly suffer any losses. The "quality factor" Q, a measure of how many oscillations take place in a cavity before damping eventually dissipates away the original excitation, would be high.

In fact, a resonant cavity would be useful whenever one would like to control radiation within a narrow frequency range. For example, all atomic transitions from one energy level to another are accompanied by the emission or absorption of photons, with energies corresponding to the difference in those energy levels. One can imagine selecting and suppressing various atomic transitions by having them take place in appropriate photonic crystals.

The important questions to address when designing a defect mode are how the defect shall be introduced into the structure, and which frequencies it will support as localized modes. First, one obvious way to introduce a defect is to allow one of the columns of the triangular lattice to grow or shrink in radius. Calling the radius of the defect column r_d, the possibilities range from $r_d = 0$, corresponding to a missing air column in the structure, to around $r_d = a = 0.75$ μm, corresponding to an air column that envelops one entire unit cell.

Next, we would like the defect to harbor modes of light that have frequencies within the band gap of the crystal. In general, to calculate defect frequencies, one needs to use a computational procedure like the one outlined in appendix D to calculate band structures for a given

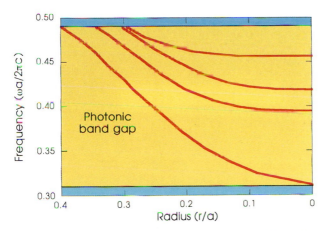

Figure 3 A plot of the TE cavity modes in a triangular lattice of air columns with $r/a = 0.45$. The photonic band gap is colored yellow. The localized modes are shown as red lines. We create the defect cavity by filling in a single air column.

$\varepsilon(\mathbf{r})$. Using this method, it is a straightforward matter to compute the defect frequencies as the defect radius varies across the entire range. The results are plotted in figure 3.

Note that it is possible not only to create a defect mode with a frequency in the band gap, but also that the defect frequency sweeps continuously across the band gap as r_d is varied. (A similar result is discussed in chapter 5.) In other words, we can "tune" the defect frequency to any value within the band gap with a judicious choice of r_d. This complete tunability is an important feature of photonic crystals—it would be analogous to the ability to tune the properties of solids by somehow adjusting the radii of single dopant atoms.

For concreteness, we choose the case $r_d = 0$, so that the triangular lattice is missing a single air column. This corresponds to the right side of figure 3. We can use our structure as a resonant cavity for modes of frequency $\omega a/2\pi c = 0.31, 0.39, 0.42$, and 0.46. In this case the defect resembles a cavity with perfectly reflecting walls, and the mode frequencies of such a cavity can be estimated just like they can for metallic cavities—by requiring that the field go to zero at the boundaries, so that either half a wavelength, a full wavelength, a wavelength and a half, etc., fits perfectly in the cavity. Of course, our more precise figures come from the results of a full computation.

Doubly-degenerate quadrupole mode

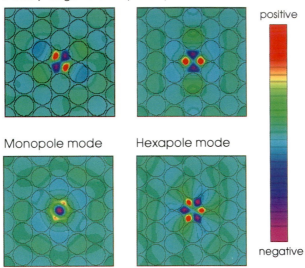

positive

negative

Monopole mode   Hexapole mode

Figure 4 The magnetic field patterns of the defect modes in a triangular lattice missing a single air column. In all cases, the color indicates the strength of the field, which points out of the page. The quadrupole, monopole, and hexapole modes correspond to the case $r_d = 0$ in figure 3 and have frequencies 0.39, 0.42, and 0.46, respectively.

The magnetic field patterns for the defect modes of the structure are shown in figure 4. As the mode frequency increases, the number of nodes (sites of zero field) increases and the mode profile becomes more complex. Note in particular the different symmetries of the defect modes. For example, one of the modes has a simple circular (monopole) symmetry and a radial node, while the higher modes have a number of lobes. Attempts to excite these modes using sources that have orthogonal symmetries would prove impossible, so the different symmetries of the defect modes might offer some additional flexibility in a resonant cavity design.

A Waveguide

We can use point defects in photonic crystals to trap light, as we have just seen. By using line defects, we can also *guide* light from one location to another. The basic idea is to carve a waveguide out of an otherwise-perfect photonic crystal. Light that propagates in the wave-guide with a frequency within the band gap of the crystal is confined to, and can be directed along, the waveguide. An example is illustrated in figure 5.

Microwaves are already commonly guided by metallic guides and coaxial cables, but we have already discussed the limitations of these methods. Visible light can be guided with fiber-optic cables, which rely on total internal reflection. However, if a fiber-optic cable takes a tight curve, the angle of incidence is too large for total internal reflection to occur, so light escapes at the corners and is lost. Photonic crystals, since they do not rely exclusively on total internal reflection, continue to confine light even around tight corners.

For these reasons, photonic-crystal waveguides can find use when-ever a monochromatic light beam needs to be guided from one location to another. These situations are becoming increasingly common in modern technology. In an optoelectronic circuit, light is guided from one end of a microchip to another. In a fiber-optic network, like those used in telecommunications, light is guided from one end of the country to another.

For variety, we will use the square lattice of rods developed in earlier chapters to illustrate this component in two dimensions. Suppose we

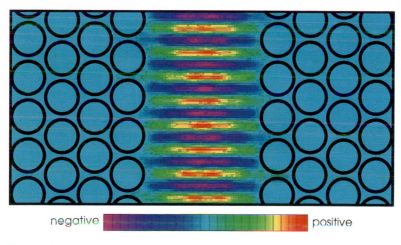

negative positive

Figure 5 Light propagating down a waveguide in a two-dimensional photonic crystal. The channel is formed by filling in rows of air columns.

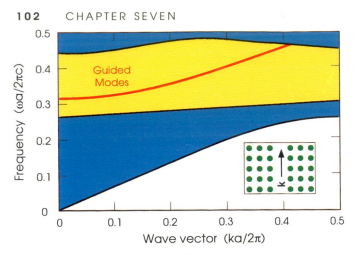

Figure 6 The projected band structure of TM modes for a waveguide in a square lattice of dielectric rods in air. The blue region contains the continuum of extended crystal states. The photonic band gap is colored yellow. The red line is the band of guided modes that runs along the waveguide. The waveguide is formed by removing one row of dielectric rods as shown in the inset.

want to construct a TM waveguide for 1.5 μm light. As before, we begin by consulting a gap map. For a square lattice of GaAs rods in air we can use figure 1 of appendix C. We find that the largest gap occurs for $r = 0.18a$. This corresponds to a mid-gap frequency near $\omega a/2\pi c = a/\lambda = 0.4$, and since $\lambda = 1.5$μm we need $a = 0.6$μm and $r = 0.1$μm. We now carve out a waveguide by removing rows of dielectric rods. For a rough estimate of the width we should choose, think of the waveguide as an empty space between two perfectly reflecting walls. For a propagating waveguide mode, the field should vanish at the walls. For the lowest-frequency mode, exactly one-half a wavelength should fit between the walls. This tells us that the width should be roughly $\lambda/2 = 0.75$ μm. We simply remove one row of dielectric rods.

To calculate the states of this system, we should do what we did in chapter 5, when we discussed the various modes associated with a surface. We can consider a straight waveguide to be a narrow channel between the surfaces of two photonic crystals. What we need are modes of the crystals, which are extended in the outside region and decay away exponentially within the crystal. Recall that we studied such states by projecting the band structure of the crystal onto the surface Brillouin zone (see the sections on surface modes in chapters 5 and 6). We also made a plot showing the four possible types of modes: extended or evanescent in the outside region, and extended or evanescent in the crystal region.

Figure 6 shows the results of this calculation for our waveguide. The blue region corresponds to states that can propagate through the crystal. The red line corresponds to guided modes, which can travel freely within the narrow waveguide channel. Note that for our waveguide configuration, there is only one guided-mode band.

Once light is induced to travel along the waveguide it really has nowhere else to go. Since the frequency of the guided mode lies within the photonic band gap, the mode is forbidden to escape into the crystal. Regardless of the shape of the waveguide, light is forced to bounce around inside it. The primary source of loss can only be reflection back out the waveguide entrance. This suggests that we may use a photonic crystal to guide light around tight corners. Returning to our square lattice of GaAs rods in air, we can carve out a waveguide with a sharp 90 degree bend as shown in figure 7. Here we plot the displacement

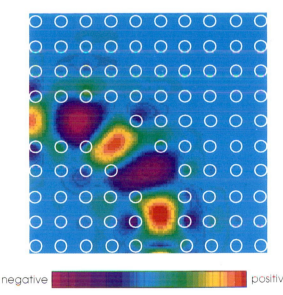

negative ▮▮▮▮▮▮▮▮▮▮▮ positive

Figure 7 The displacement field of a TM mode traveling around a sharp bend in a waveguide carved out of a square lattice of dielectric rods. Light is coming in from the bottom and exiting at the left.

field of a propagating TM mode[5] as it travels around the corner. Even though the radius of curvature of the bend is less than the wavelength of the light, very nearly all the light that goes in one end comes out the other!

Epilogue

Throughout most of the text, we have emphasized the basic physical principles underlying photonic crystals. By examining a few basic components in detail, we hoped to suggest the practical importance and versatility of photonic-crystal technology. We believe that these examples are only the tip of an iceberg of possibilities that are yet to be discovered. Our goal in writing this text was to stimulate the imaginations of researchers in diverse fields who might benefit from the possibilities afforded by photonic crystals and turn them into realities.

[5]Note that a TM mode defined with respect to the crystal plane normal corresponds to a conventional TE waveguide mode (Jackson 1962) defined with respect to the direction of propagation.

Appendixes

A

Comparisons with Quantum Mechanics

Throughout the text, especially in chapters 2 and 3, we make several comparisons between our formalism and the equations of quantum mechanics and solid-state physics. In this appendix, we present an extensive listing of these comparisons. It will hopefully serve as both a brief summary of the phenomena surrounding photonic crystals, and a way to relate them to (perhaps) familiar concepts in other fields.

The heart of the subject of photonic crystals is the propagation of electromagnetic waves in a periodic dielectric medium. In a sense, quantum mechanics is also the study of wave propagation, although the waves are a bit more abstract. At the atomic scale, particles (like the electron) begin to display wavelike properties, including interference and nonlocalization. The function that contains the information about the particle obeys the Schrödinger equation, which bears some resemblance to a familiar wave equation.

It therefore comes as no surprise that the study of quantum mechanics in a periodic potential contains direct parallels to our study of electromagnetism in a periodic dielectric. Since the quantum mechanics of periodic potentials is the basic theory of solid-state physics, the field of photonic crystals can also inherit some of the theorems and terminology of solid-state physics, in slightly modified form. Table 1 lists some of these correspondences.

Table 1 Extended comparison between quantum mechanics in a periodic potential and electromagnetism in a periodic dielectric

	QUANTUM MECHANICS IN A PERIODIC POTENTIAL (CRYSTAL)	ELECTROMAGNETISM IN A PERIODIC DIELECTRIC (PHOTONIC CRYSTAL)
What is the "main function" that contains all of the information?	The scalar wave function $\Psi(\mathbf{r}, t)$.	The magnetic vector field $\mathbf{H}(\mathbf{r}, t)$.
How do we separate out the time dependence of the function (into normal modes)?	$\Psi(\mathbf{r}, t) = \sum_E c_E \Psi_E(\mathbf{r}) e^{iEt/\hbar}$ Expand in a set of energy eigenstates $\Psi_E(\mathbf{r})$	$\mathbf{H}(\mathbf{r}, t) = \sum_\omega c_\omega \mathbf{H}_\omega(\mathbf{r}) e^{i\omega t}$ Expand in a set of harmonic modes $\mathbf{H}_\omega(\mathbf{r})$.
What is the "master equation" that determines the normal modes of the system?	$\left(\dfrac{p^2}{2m} + V(\mathbf{r}) \right) \Psi_E(\mathbf{r}) = E \Psi_E(\mathbf{r})$ The Schrödinger equation.	$\nabla \times \dfrac{1}{\varepsilon(\mathbf{r})} \nabla \times \mathbf{H}_\omega(\mathbf{r}) = \dfrac{\omega^2}{c^2} \mathbf{H}_\omega(\mathbf{r})$ The Maxwell equations.
Are there any other conditions on the main function?	Yes, it must be normalizable.	Yes, the field must be both normalizable and transverse: $\nabla \cdot \mathbf{H} = 0$.
Where does the periodicity of the system enter?	The potential: $V(\mathbf{r}) = V(\mathbf{r} + \mathbf{R})$, for all lattice vectors R.	The dielectric: $\varepsilon(\mathbf{r}) = \varepsilon(\mathbf{r} + \mathbf{R})$, for all lattice vectors R.
Is there any interaction between normal modes?	Yes, there is an electron-electron repulsive interaction that makes large-scale computation difficult.	In the linear regime, light modes can pass right through one another undisturbed, and can be calculated independently.

What important properties do the normal modes have in common?

Eigenstates with different energies are orthogonal, they have real eigenvalues, and can be found with a variational principle.

Modes with different frequencies are orthogonal, they have real positive eigenvalues, and can be found with a variational principle.

On what fact about the master equation do the important properties rely?

The Hamiltonian H is a linear, Hermitian operator.

$$E_{var} = \frac{\langle \Psi | H | \Psi \rangle}{\langle \Psi | \Psi \rangle}$$

The Maxwell operator Θ is a linear, Hermitian operator.

$$E_{var} = \frac{(\mathbf{H}, \Theta \mathbf{H})}{(\mathbf{H}, \mathbf{H})}$$

What is the variational theorem we use to determine the normal modes and frequencies?

E_{var} is minimized when Ψ is an eigenstate of H.

E_{var} is minimized when $\mathbf{H}$ is a normal mode of Θ.

What is the heuristic that goes along with the variational theorem?

The wave function concentrates in regions of low potential, while remaining orthogonal to lower states.

The fields concentrate their electrical energy in high-ε regions, while remaining orthogonal to lower modes.

What is the physical energy of the system?

The eigenvalue E of the Hamiltonian.

$$E = \left(\frac{1}{8\pi}\right) \iint d\mathbf{r} \left(\frac{1}{\varepsilon}|\mathbf{D}|^2 + |\mathbf{H}|^2\right)$$

The electromagnetic energy.

Is there a natural length scale to the system?

Usually, because constants such as the Bohr radius set the length scale.

No, solutions are scalable to any length scale.

What is the mathematical statement that says: "A is a symmetry of the system"?

A commutes with the Hamiltonian: $[A, H] = 0$.

A commutes with the Maxwell operator: $[A, \Theta] = 0$.

Table 1 Continued

	Quantum Mechanics in a Periodic Potential (Crystal)	Electromagnetism in a Periodic Dielectric (Photonic Crystal)
How do we classify the normal modes using a system's symmetries?	Distinguish them by how they transform under a symmetry operation A.	Distinguish them by how they transform under a symmetry operation A.
What does the discrete translational symmetry of a crystal allow us to do?	$\Psi_{\mathbf{k}}(\mathbf{r}) = u_{\mathbf{k}}(\mathbf{r})e^{i\mathbf{k}\cdot\mathbf{r}}$ Write the wave function in Bloch form.	$\mathbf{H}_{\mathbf{k}}(\mathbf{r}) = \mathbf{u}_{\mathbf{k}}(\mathbf{r})e^{i\mathbf{k}\cdot\mathbf{r}}$ Write the harmonic modes in Bloch form.
What are the allowable values for the wave vvector $\mathbf{k}$*?*	They lie in the Brillouin zone in reciprocal space.	They lie in the Brillouin zone in reciprocal space.
What do we mean by the term "band structure"?	The functions $E_n(\mathbf{k})$, which tell us the energies of the allowed eigenstates.	The functions $\omega_n(\mathbf{k})$, which tell us the frequencies of the allowed harmonic modes.
What is the physical origin of the band structure?	The electron wave scatters coherently from the different potential regions.	The electromagnetic fields scatter coherently at the interfaces between different dielectric regions.
What happens inside a gap in the band structure?	No propagating electrons in that energy range are allowed to exist, regardless of wave vector.	No extended modes in that frequency range are allowed to exist, regardless of wave vector.
What do we call the bands immediately above and below the gap?	The band above the gap is the *conduction band;* the band below the gap is the *valence band.*	The band above the gap is the *air band;* the band below the gap is the *dielectric band.*

How do we introduce defects into the system?	By adding foreign atoms to the crystal, which breaks the translational symmetry of the atomic potential.	By changing the dielectric constant of certain regions, which breaks the translational symmetry of $\varepsilon(\mathbf{r})$.
What is the result of introducing a defect?	It might create an allowed state in a band gap, thereby permitting a localized electron state around the defect.	It might create an allowed state in a band gap, thereby permitting a localized mode around the defect.
How do we classify different types of defects?	Donor atoms pull states from the conduction band into the gap; acceptor atoms pull states from the valence band into the gap.	Dielectric defects pull states from the air band into the gap; air defects pull states from the dielectric band into the gap.
In short, why is the study of the physics of the system important?	We can tailor the *electronic* properties of materials to our needs.	We can tailor the *optical* properties of materials to our needs.

B

The Reciprocal Lattice and
the Brillouin Zone

Beginning in chapter 4, we used Bloch's theorem to express an electromagnetic mode as a plane wave that is modulated by a periodic function $\mathbf{u}(\mathbf{r})$. The function $\mathbf{u}$ shares the same periodicity as the crystal. We also argued that we need only consider wave vectors $\mathbf{k}$ that lie in a certain region of the reciprocal lattice known as the *Brillouin zone*.

This is not news to anyone who has studied solid-state physics or other fields in which lattices play a large role. But for readers who have never encountered the Brillouin zone, this appendix will provide enough information to completely understand the material in this text. Specifically, we will introduce the reciprocal lattice and identify the Brillouin zone for some simple lattices that we use throughout the text. In addition, we describe the Miller index notation we use in chapter 6 to refer to crystal planes. For more details, it is best to consult the first few chapters of a solid-state physics text, such as Kittel (1986) or Ashcroft and Mermin (1976).

The Reciprocal Lattice

Suppose we have a function $f(\mathbf{r})$ that is periodic on a lattice; that is, suppose $f(\mathbf{r}) = f(\mathbf{r} + \mathbf{R})$ for all vectors $\mathbf{R}$ that translate the lattice into itself (i.e., connect one lattice point to the next). Our dielectric function $\varepsilon(\mathbf{r})$ is an example of such a function. The set of vectors $\mathbf{R}$ is called the *lattice vectors*.

A natural thing to do when analyzing periodic functions is to take the Fourier transform; that is, we build the periodic function $f(\mathbf{r})$ out of plane waves with various wave vectors. The expansion looks like this:

$$f(\mathbf{r}) = \int d\mathbf{q}\, g(\mathbf{q})e^{i\mathbf{q}\cdot\mathbf{r}}. \tag{1}$$

Here $g(\mathbf{q})$ is the coefficient on the plane wave with wave vector $\mathbf{q}$. An expansion like this can be performed on *any* well-behaved function. But our function f is periodic on the lattice—what information does this tell us about the expansion? Requiring $f(\mathbf{r}) = f(\mathbf{r} + \mathbf{R})$ in the expansion yields

$$f(\mathbf{r} + \mathbf{R}) = \int d\mathbf{q}\, g(\mathbf{q})e^{i\mathbf{q}\cdot\mathbf{r}}e^{i\mathbf{q}\cdot\mathbf{R}} = f(\mathbf{r}) = \int d\mathbf{q}\, g(\mathbf{q})e^{i\mathbf{q}\cdot\mathbf{r}}.$$

The periodicity of f tells us that its Fourier transform $g(\mathbf{q})$ has the special property $g(\mathbf{q}) = g(\mathbf{q})\exp(i\mathbf{q}\cdot\mathbf{R})$. But this is impossible, unless either $g(\mathbf{q}) = 0$ or $\exp(i\mathbf{q}\cdot\mathbf{R}) = 1$. In other words, the transform $g(\mathbf{q})$ is zero everywhere, except for spikes at the values of $\mathbf{q}$ such that $\exp(i\mathbf{q}\cdot\mathbf{R}) = 1$ for all $\mathbf{R}$.

What we have just discovered is that if we are building a lattice-periodic function f out of plane waves, we need only use those plane waves with wave vectors $\mathbf{q}$ such that $\exp(i\mathbf{q}\cdot\mathbf{R}) = 1$ for all of the lattice vectors $\mathbf{R}$. This statement has a simple analog in one dimension: if we are building a function $f(x)$ with period τ out of sinusoids, then we need only use the "fundamental" sinusoid with period τ and its "harmonics" with periods $\tau/2$, $\tau/3$, $\tau/4$ and so on.

Those vectors $\mathbf{q}$ such that $\exp(i\mathbf{q}\cdot\mathbf{R}) = 1$, or, equivalently, $\mathbf{q}\cdot\mathbf{R} = n2\pi$, are called *reciprocal lattice vectors* and are usually designated by the letter $\mathbf{G}$. They form a lattice of their own; for example, adding two reciprocal lattice vectors $\mathbf{G}_1$ and $\mathbf{G}_2$ yields another reciprocal lattice vector, as you can easily verify. We can build our function $f(\mathbf{r})$ with an appropriate weighted sum over all of the reciprocal lattice vectors, as follows:

$$f(\mathbf{r}) = \sum_{\mathbf{G}} f_{\mathbf{G}}\, e^{i\mathbf{G}\cdot\mathbf{r}}.$$

Constructing the Reciprocal Lattice Vectors

Given a lattice with a set of lattice vectors $\{\mathbf{R}\}$, how can we determine all of the reciprocal lattice vectors $\{\mathbf{G}\}$? We need to find all $\mathbf{G}$ such that $\mathbf{G}\cdot\mathbf{R}$ is some integral multiple of 2π for every $\mathbf{R}$.

We know that every lattice vector $\mathbf{R}$ can be written in terms of the *primitive lattice vectors,* which are the smallest vectors pointing from one lattice point to another. For example, on a simple cubic lattice with spacing a, the vectors $\mathbf{R}$ would all be of the form $\mathbf{R} = la\hat{\mathbf{x}} + ma\hat{\mathbf{y}} + na\hat{\mathbf{z}}$, where (l, m, n) are integers. In general, we call the primitive lattice vectors $\mathbf{a}_1$, $\mathbf{a}_2$, and $\mathbf{a}_3$. They need not be of unit length.

We have already mentioned that the reciprocal lattice vectors $\{\mathbf{G}\}$ form a lattice of their own. In fact, the reciprocal lattice has a set of primitive vectors $\mathbf{b}_i$ as well, so that every reciprocal lattice vector $\mathbf{G}$ can be written as $\mathbf{G} = l\,\mathbf{b}_1 + m\,\mathbf{b}_2 + n\,\mathbf{b}_3$. Our requirement that $\mathbf{G} \cdot \mathbf{R} = n2\pi$ boils down to the primitive requirement:

$$\mathbf{G} \cdot \mathbf{R} = (l\mathbf{a}_1 + m\mathbf{a}_2 + n\mathbf{a}_3) \\ \cdot (l'\mathbf{b}_1 + m'\mathbf{b}_2 + n'\mathbf{b}_3) = N2\pi. \tag{4}$$

For all choices of (l, m, n), the above must hold for some N. A little thought will suggest that we could satisfy the above if we construct the $\mathbf{b}_i$ so that $\mathbf{a}_i \cdot \mathbf{b}_j = 2\pi$ if $i = j$, and 0 if $i \neq j$. More compactly, we write $\mathbf{a}_i \cdot \mathbf{b}_j = 2\pi\delta_{ij}$. Given the set $\{\mathbf{a}_1, \mathbf{a}_2, \mathbf{a}_3\}$, our task is to find the corresponding set $\{\mathbf{b}_1, \mathbf{b}_2, \mathbf{b}_3\}$ such that $\mathbf{a}_i \cdot \mathbf{b}_j = 2\pi\delta_{ij}$.

One way to do this is to exploit a feature of the cross product. Remembering that $\mathbf{x} \cdot (\mathbf{x} \times \mathbf{y}) = 0$ for any vectors $\mathbf{x}$ and $\mathbf{y}$, we can construct the primitive reciprocal lattice vectors with the following recipe:

$$\mathbf{b}_1 = 2\pi \frac{\mathbf{a}_2 \times \mathbf{a}_3}{\mathbf{a}_1 \cdot \mathbf{a}_2 \times \mathbf{a}_3} \quad \mathbf{b}_2 = 2\pi \frac{\mathbf{a}_3 \times \mathbf{a}_1}{\mathbf{a}_1 \cdot \mathbf{a}_2 \times \mathbf{a}_3}$$
$$\mathbf{b}_3 = 2\pi \frac{\mathbf{a}_1 \times \mathbf{a}_2}{\mathbf{a}_1 \cdot \mathbf{a}_2 \times \mathbf{a}_3}. \tag{5}$$

In summary, when we take the Fourier transform of a function that is periodic on a lattice, we only need to include terms with wave vectors that are reciprocal lattice vectors. To construct the reciprocal lattice vectors, we take the primitive lattice vectors and perform the operations of (5).

The Brillouin Zone

In chapter 3, we found that the discrete translational symmetry of a photonic crystal allows us to classify the electromagnetic modes with a wave vector $\mathbf{k}$. The modes can be written in "Bloch form," consisting

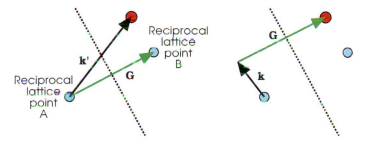

Figure 1 Characterization of the Brillouin zone. The dotted line is the perpendicular bisector of the line joining two lattice points (blue). If we choose the left point as the origin, any lattice vector (such as **k′**) that reaches to an arbitrary point on the other side (red) can be expressed as the sum of a same-side vector (such as **k**) plus a reciprocal lattice vector **G**.

of a plane wave modulated by a function that shares the periodicity of the lattice:

$$\mathbf{H_k(r)} = e^{i\,\mathbf{k\cdot r}}\mathbf{u_k(r)} = e^{i\,\mathbf{k\cdot r}}\mathbf{u_k(r + R)}. \tag{6}$$

One important feature of the Bloch states is that different values of **k** do not necessarily lead to different modes. Specifically, a mode with wave vector **k** and a mode with wave vector **k** + **G** are the same mode, if **G** is a reciprocal lattice vector. The wave vector **k** serves to specify the phase relationship between the various cells that are described by **u**. If **k** is incremented by **G**, then the phase between cells is incremented by **G · R**, which we know is $n2\pi$ and not really a phase difference at all. So incrementing **k** by **G** results in the same physical mode.

This means that there is a lot of redundancy in the label **k**. We can restrict our attention to a finite zone in reciprocal space in which you *cannot* get from one part of the volume to another by adding any **G**. All values of **k** that lie outside of this zone, by definition, can be reached from within the zone by adding **G**, and are therefore redundant labels.

This zone is the *Brillouin zone*. A more visual way to characterize it is the following: around any lattice point in reciprocal space, highlight the volume which is closer to *that* lattice point than to *any other* lattice point. If we call the original lattice point the origin, then the highlighted region is the Brillouin zone.

The two definitions are equivalent. If a particular **k** is closer to a neighboring lattice point, you can always reach it by staying close to the original lattice vector and then translating by the **G** that reaches from one lattice point to the other. This situation is depicted in figure 1.

The next few sections will be devoted to a study of the reciprocal lattice vectors and Brillouin zones of some of the particular lattices that appear in this text.

Two-Dimensional Lattices

In chapter 5, we worked extensively with photonic crystals that are based on a square or triangular lattice. What are the reciprocal lattice vectors and Brillouin zones of each?

For a square lattice with spacing a, the lattice vectors are $\mathbf{a}_1 = a\hat{\mathbf{x}}$ and $\mathbf{a}_2 = a\hat{\mathbf{y}}$. To use our prescription (5), we can use a third basis vector in the z-direction of any length, since the crystal is homogeneous in that direction. The results are $\mathbf{b}_1 = (2\pi/a)\hat{\mathbf{y}}$ and $\mathbf{b}_2 = (2\pi/a)\hat{\mathbf{x}}$. The reciprocal lattice is also a square lattice, but with spacing $2\pi/a$ instead of a. The name "reciprocal lattice" suits this fact well.

To determine the Brillouin zone, we fix our attention on a particular lattice point (the origin) and shade the area that is closer to that point than to any other. Geometrically, we draw perpendicular bisectors of every lattice vector that starts at the origin. Each bisector divides the lattice into two half-planes, one of which contains our lattice point. The intersection of all the half-planes that contain our lattice point is

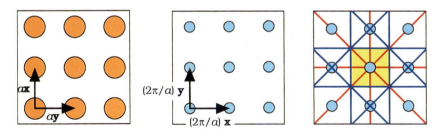

Figure 2 The square lattice. On the left is the network of lattice points in real space. In the middle is the corresponding reciprocal lattice. On the right is the construction of the Brillouin zone: taking the center point as the origin, we draw the lines connecting the origin to other lattice points (red), their perpendicular bisectors (blue), and highlight the square boundary of the Brillouin zone (yellow).

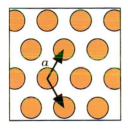
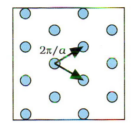
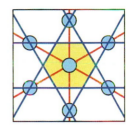

Figure 3 The triangular lattice. On the left is the network of lattice points in real space. In the middle is the corresponding reciprocal lattice, which in this case is a rotated version of the original. On the right is the Brillouin zone construction. In this case, the Brillouin zone is a hexagon centered around the origin.

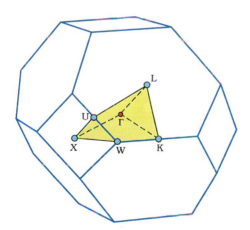

Figure 4 The Brillouin zone for the face-centered cubic lattice. The reciprocal lattice is a body-centered cubic lattice, and the Brillouin zone is a truncated octahedron with center at Γ. Also shown are some of the names which are traditionally given to the special directions in the zone. The irreducible Brillouin zone is the yellow polyhedron with corners at Γ, X, U, L, W, and K.

the Brillouin zone. The square lattice vectors, its reciprocal lattice vectors, and its Brillouin zone are shown in figure 2.

We can handle the triangular lattice in a similar fashion. The lattice vectors are $a(\hat{\mathbf{y}} + \hat{\mathbf{x}}\sqrt{3})/2$ and $a(\hat{\mathbf{y}} - \hat{\mathbf{x}}\sqrt{3})/2$, as shown in figure 3. Using our prescription (5), we obtain reciprocal lattice vectors $(2\pi/a)$ $(\hat{\mathbf{x}}\sqrt{3} + \hat{\mathbf{y}})/2$ and $(2\pi/a)(\hat{\mathbf{x}}\sqrt{3} - \hat{\mathbf{y}})/2$. This is again a triangular lattice, but rotated by $90°$ with respect to the first, and with spacing $2\pi/a$. The Brillouin zone, as determined by the construction outlined above, is a hexagon.

Three-Dimensional Lattices

Our construction works in three dimensions as well, with the proper extensions. Instead of drawing perpendicular bisectors to the lattice vectors, draw perpendicular *planes* that bisect the line. Then, the Brillouin zone is the intersection of all of the half-spaces that contain the origin.

The one lattice we devote the most attention to in the text is the face-centered cubic lattice. This is like a cubic lattice, but with an additional lattice point at each face of the cubes. The primitive lattice vectors are $\mathbf{a}_1 = a/2(\hat{\mathbf{x}} + \hat{\mathbf{y}})$, $\mathbf{a}_2 = a/2(\hat{\mathbf{y}} + \hat{\mathbf{z}})$ and $\mathbf{a}_3 = a/2(\hat{\mathbf{x}} + \hat{\mathbf{z}})$.

Using our recipe (5), we find that the reciprocal lattice vectors form a *body*-centered cubic lattice, which is like a cubic lattice with an additional lattice point at the *center* of every cube. The primitive reciprocal lattice vectors are $\mathbf{b}_1 = (2\pi/a)(\hat{\mathbf{x}} + \hat{\mathbf{y}} - \hat{\mathbf{z}})$, $\mathbf{b}_2 = (2\pi/a)(-\hat{\mathbf{x}} + \hat{\mathbf{y}} + \hat{\mathbf{z}})$, and $\mathbf{b}_3 = (2\pi/a)(\hat{\mathbf{z}} + \hat{\mathbf{x}} - \hat{\mathbf{y}})$.

The Brillouin zone, which can be determined in the usual fashion, is a truncated octahedron. It is depicted in figure 4, along with the traditional names for the special points in the irreducible Brillouin zone.

Miller Indices

We return to the real lattice, as opposed to the reciprocal lattice, for one last miscellaneous topic. It is often convenient to have a systematic way of referring to directions and planes in a crystal. For example, in chapter 6 we refer to various cross sections and directions in Yablonovite. The traditional way to refer to planes in a crystal lattice is the system of Miller indices. To characterize a plane, we need to specify

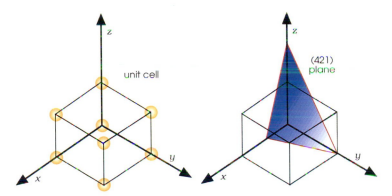

Figure 5 The Miller indices for a crystal plane. *Left:* A plot of the unit cell. The length of the unit cell in each direction is the lattice constant for that direction. *Right:* The shaded plane is named by locating the intercepts of the plane with the axes, as described in the text.

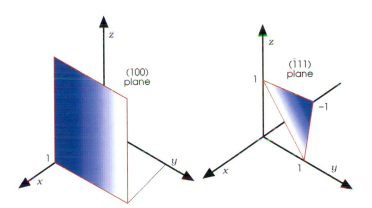

Figure 6 Special cases of the Miller index system. *Left:* The (100) plane of a simple cubic lattice. The plane never intersects the y- or z-axes, so the indices are zero. *Right:* The ($\bar{1}11$) plane of the simple cubic lattice. The plane intersects the x-axis at −1, so a bar is placed over the corresponding index.

three noncollinear points in the plane. The Miller indices of a crystal plane are integers that give the location of three such points, relative to the lattice vectors of the crystal.

Suppose the crystal has lattice vectors $\mathbf{a}_1$, $\mathbf{a}_2$, and $\mathbf{a}_3$. The unit cell of the crystal can be plotted with $\mathbf{a}_1$, $\mathbf{a}_2$, and $\mathbf{a}_3$ along its edges. To name a particular plane in the crystal, we draw the plane on such a plot; in general, the plane will intersect each of the axes $\mathbf{a}_1$, $\mathbf{a}_2$, and $\mathbf{a}_3$. An example of such a plot is shown in figure 5, for the special case $\mathbf{a}_1 = a\hat{\mathbf{x}}$, $\mathbf{a}_2 = a\hat{\mathbf{y}}$, $\mathbf{a}_3 = a\hat{\mathbf{z}}$. We would like to name a surface that intersects the x-axis at one-half lattice constant away from the origin, the y-axis one lattice constant away, and the z-axis two lattice constants away.

First, we take the reciprocals of these intercepts: 2, 1, and $\frac{1}{2}$. Next, we find the smallest integers that have the same ratios as these three: 4, 2, and 1. These integers are the Miller indices, and we refer to this plane as the (421) plane. The direction perpendicular to this plane is known as the [421] direction.

There are two common complications. First, if the surface is parallel to a particular lattice vector, then it will never intercept that axis. The intercept is infinite, so to speak, and the reciprocal is zero. The Miller index for that direction is zero. figure 6 shows an example of this case, for the (100) plane of a cubic lattice.

Second, if the plane intercepts an axis at a negative value, the Miller index for that direction is negative. Traditionally, this is indicated by placing a bar over a positive integer instead of using a minus sign. For example, the ($\bar{1}$11) plane of the cubic lattice is shown in figure 6.

C

Atlas of Band Gaps
for Two-Dimensional
Photonic Crystals

In this appendix, we will chart the locations of the photonic band gaps in several different two-dimensional photonic crystals. We hope this will serve two purposes. First, to demonstrate that the engineering possibilities with photonic crystals are enormous; the class of possible photonic crystals is so large that designing crystals with gaps in desirable locations is a real possibility. Second, to serve as a convenient atlas of easily fabricated crystals, in case one of these crystals suits a particular application.

Reading a Gap Map

A plot of the locations of the photonic band gaps of a crystal, as one or more of the parameters of the crystal are varied, is what we call a "gap map." In this appendix we will present the gap maps for the square and triangular lattices of cylindrical columns as the radius of the columns varies. We will do this both for dielectric columns in air, and for air columns in dielectric, both with a dielectric contrast of 11.4.[1] We will plot the TE and TM polarizations together. A larger compilation of gap maps, for a variety of different lattices and index contrasts, can be found in Meade et al. (1993b).

Along the horizontal axis of a gap map is the radius of the columns; along the vertical axis is the frequency (in dimensionless units). The locations of the band gaps are outlined for both the TE and TM

---

[1]This is the dielectric contrast between gallium arsenide (GaAs) and air, for wavelengths between 1.0 μm and 10.0 μm, as reported in Pankove (1971).

polarizations. To use a gap map to find a crystal for a specific application, the frequency and the lattice constant must be scaled to the desired levels, using the scaling laws developed in chapter 2.

A Guided Tour of the Gap Maps

The first two-dimensional photonic crystal that we will consider consists of parallel columns arranged in a square lattice. The columns have radius r, and the lattice constant is a. We first consider the case of dielectric ($\varepsilon = 11.4$) columns in air. The gap map is shown in figure 1.

At a glance, the gap map reveals some interesting regularities. First, the gaps all decrease in frequency as r/a increases. This is an expected feature, since the frequency scales as $1/\sqrt{\varepsilon}$ in a medium of dielectric constant ε; and as r/a increases, the average dielectric constant of the medium steadily increases. Second, even though the plot extends all the way to $r/a = 0.70$, for which the dielectric columns fill space, all of the gaps seal up by the time $r/a = 0.50$. At that value, the dielectric columns begin to touch one another. The third and perhaps most remarkable feature is the repetition of the lowest, largest gap at higher frequencies; progressively smaller copies of the lowest gap are stacked above it at roughly equal intervals. For $r/a = 0.38$, there are four TM gaps in the band structure!

But for the TE polarization, figure 1 shows a much more barren

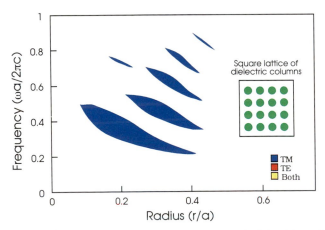

Figure 1 Gap map for a square lattice of dielectric columns, $\varepsilon = 11.4$.

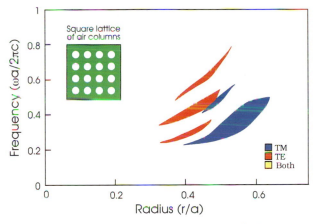

Figure 2 Gap map for a square lattice of air columns drilled in a dielectric medium, $\varepsilon = 11.4$.

terrain. There are no significant TE gaps at all for the square lattice in the frequency range displayed. Our results are in agreement with the heuristic of chapter 5, which holds that connectivity of high-ε regions is conducive to TE gaps, and isolated patches of high-ε regions lead to TM gaps.

Now we reverse the dielectric configuration, so that the columns of the square lattice have $\varepsilon = 1$, and the surrounding medium has $\varepsilon = 11.4$. The gap map for both polarizations is shown in figure 2. Immediately we see that frequencies increase with r/a this time, since the average dielectric constant increases as the air columns grow. The gap structure for the TM modes seems to open up around $r/a = 0.45$, in contrast with the sharp cutoff seen for the dielectric columns. Apparently, connectivity between the air columns is of importance, since figure 2 displays a significant change of behavior around $r/a = 0.5$.

For the TE polarization, the square lattice of air columns fares a bit better than that of dielectric columns, as figure 2 attests. Several thin gaps may be noted. However, none of these gaps overlaps a gap for the TM polarization, so there is no *complete* band gap for the square lattice for this value of the dielectric contrast. We shall see that our next structure, the triangular lattice of columns, does possess such a complete band gap.

Our second structure, the triangular lattice of columns, is depicted in the inset in figure 3. Here the columns begin touching one another

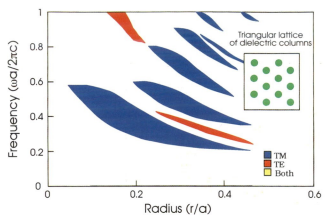

Figure 3 Gap map for a triangular lattice of dielectric columns, $\varepsilon = 11.4$.

at $r/a = 0.50$, and fill space at $r/a = 0.58$. Again we can distinguish the cases of dielectric ($\varepsilon = 11.4$) columns in air, and air columns in dielectric. We begin with dielectric columns in air.

Figure 3 displays the gap map. The remarkable self-similarity of the figure 1, which was for the TM polarization of the square lattice of dielectric columns, is mirrored here. The successive gaps are similar in shape and orientation, and stack regularly upon one another. The cutoff at $r/a = 0.45$ is once again near the column-touching condition.

The gap map for the TE polarization is almost as sparse as the corresponding case of the square lattice. Only a few slivers are noticeable. The gross properties of the map (decrease in ω with r/a, transition at $r/a = 0.5$) follow the same trends as those already discussed.

Finally, we turn to the case of air columns in dielectric. The gap map is shown in figure 4. Although the gaps are hardly comparable in size to the gaps for dielectric columns, the lowest gap should be noted. It happens to occur at the same location as the lowest TE gap, thereby forming a complete band gap.

The enormous TE gap between $r/a = 0.20$ and 0.50 provides ample space for an overlap with the TM gap noted above. Thus, for r/a around 0.45, the triangular lattice of air columns possesses a complete band gap for all polarizations for frequencies around $0.45(2\pi c/a)$. This discovery was first reported in Meade et al. (1992) and Villeneuve and Piche (1992).

Now that the survey is complete, we can assemble the highlights

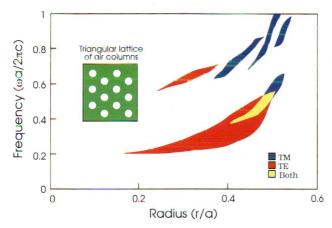

Figure 4 Gap map for a triangular lattice of air columns drilled in a dielectric medium, $\varepsilon = 11.4$.

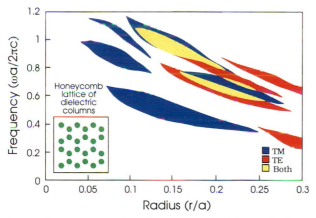

Figure 5 Gap map for a honeycomb lattice of dielectric columns, $\varepsilon = 11.4$.

from our atlas of the square and triangular lattices. For dielectric columns in air, band gaps are most abundant for the TM polarization, as long as the columns are large enough to touch one another. For air columns, the TE polarization has more band gaps, and there is even a complete band gap for the triangular lattice.

Although band gaps appear in such simple structures as the square and triangular lattices, these gaps are not optimal. In particular, it may be desirable to have a two-dimensional structure that has an overlapping band gap for both polarizations and is easy to fabricate. Although we saw in figure 4 that the triangular lattice of air columns has such a gap, it occurs at a diameter of $d = 0.95a$, at a midgap frequency of $\omega a/2\pi c = 0.48$. Thus, this structure has very thin dielectric veins of width $0.05a$ between the air columns. In fact, to fabricate such a structure with a band gap at $\lambda = 1.5$ μm would require a feature size of 0.035 μm. While such fine feature sizes may be fabricated using e-beam lithography as in figure 2 of chapter 7, this is a very difficult procedure (see Wendt et al. 1993).

Luckily, there are so many possibilities for the geometry of the lattice that suitable structures can be found. Consider, for instance, the "honeycomb lattice." The gap map for this structure is presented in figure 5. This figure shows a large overlap around $r/a = 0.14$ and $\omega a/2\pi c \sim 1.0$, which is of much larger extent than the complete band gap of the triangular lattice. To fabricate such a structure with a band gap at $\lambda = 1.5$ μm would require a feature size of 0.45 μm. This improvement should make the production of such two-dimensional lattices less formidable.

D

Computing Photonic Band Structures

Ab Initio Calculations

In the text, we present the band structures for a number of different photonic crystals and explain the interesting features of each. But how does one determine the band structure in the first place? Given a photonic crystal $\varepsilon(\mathbf{r})$, how does one obtain the band structure functions $\omega_n(\mathbf{k})$?

One alternative is to manufacture a large crystal and perform scattering experiments. By irradiating the crystal with a beam of light, and determining which values of $\Delta\mathbf{k}$ are allowed at a given frequency ω, an experimentalist can fill in the band structure. Of course, this is easier said than done; the fabrication process and experimental details are quite intricate.

But one exciting aspect of this field is that the band structures can be calculated *ab initio* (from first principles), and the results of the computations have consistently been in perfect agreement with experiments that have been possible. The Maxwell equations are practically exact. We do not need to make any questionable assumptions or simplifications, as is usually the case in computer simulations. With the appropriate theoretical tools, we can design photonic crystals with desirable properties on a computer, and *then* manufacture them! The computer becomes the pre-laboratory.

In this section we will briefly outline the computational scheme that we allude to in the text at several points. We have employed this scheme in our research, and have successfully compared its results to the results of experimenters. We used it in creating the band structures in this text. For a more thorough discussion, see Meade et al. (1993a). For a description of some other methods of calculating band structures, see Ho et al. (1990) and Sözüer et al. (1992).

Computational Scheme

The goal is to solve the "master equation" for the magnetic modes of the photonic crystal, subject to the transversality requirement. Doing so allows us to determine which are the allowed mode frequencies for a given crystal, and which wave vectors **k** are associated with those modes. In other words, we can determine the band structure. Recapitulating the results of chapter 2, the master equation is

$$\nabla \times \left(\frac{1}{\varepsilon(\mathbf{r})} \nabla \times \mathbf{H}_\omega(\mathbf{r}) \right) = \left(\frac{\omega}{c} \right)^2 \mathbf{H}_\omega(\mathbf{r}). \tag{1}$$

In addition, the transversality requirement forces $\nabla \cdot \mathbf{H}_\omega(\mathbf{r}) = 0$. This time we have included the subscript "ω" to emphasize that the field pattern corresponds to a specific frequency.

Next we expand the field pattern into a set of plane waves. This will convert the differential equation into a system of linear equations that we can readily solve on a computer. Requiring that **H(r)** shares the periodicity of $\varepsilon(\mathbf{r})$ amounts to only including the crystal's reciprocal lattice vectors **G** in the expansion:

$$\mathbf{H}_\omega^\mathbf{k}(\mathbf{r}) = \sum_{\mathbf{G}\lambda} h_{\mathbf{G}\lambda} \hat{\mathbf{e}}_\lambda e^{[i(\mathbf{k}+\mathbf{G})\cdot\mathbf{r}]}. \tag{2}$$

We have identified each mode with a wave vector **k**, and each mode is built out of plane waves with wave vector **k** + **G** for all reciprocal lattice vectors **G**. The polarization of each plane wave is one of the two unit vectors $\hat{\mathbf{e}}_\lambda$, indexed by the label λ. The transversality requirement forces us to consider only plane waves with $\hat{\mathbf{e}}_\lambda \cdot (\mathbf{k} + \mathbf{G}) = 0$.

Before inserting this expansion into the master equation, we should also expand the dielectric function $\varepsilon(\mathbf{r})$ in plane waves. Again, we need only consider plane waves whose wave vectors are reciprocal lattice vectors, because of the periodicity. Call $\varepsilon(\mathbf{G}, \mathbf{G}')$ the coefficient on the plane wave with wave vector $(\mathbf{G}' - \mathbf{G})$. Inserting both expansions into the master equation (1), we obtain a system of linear equations on the expansion coefficients:

$$\sum_{(\mathbf{G}\lambda)'} \Theta^\mathbf{k}_{(\mathbf{G}\lambda),(\mathbf{G}\lambda)'} h_{(\mathbf{G}\lambda)} = \left(\frac{\omega}{c} \right)^2 h_{(\mathbf{G}\lambda)}. \tag{3}$$

Here we have identified the **k**-dependent matrix Θ as the following quantity:

$$\Theta^k_{(G\lambda),(G\lambda)'} = [(\mathbf{k} + \mathbf{G}) \times \hat{\mathbf{e}}_\lambda] \cdot [(\mathbf{k} + \mathbf{G}') \times \hat{\mathbf{e}}_{\lambda'}]\varepsilon^{-1}(\mathbf{G}, \mathbf{G}'). \quad (4)$$

From here, we use one of the many techniques available to solve Hermitian eigenvalue problems.[1] We saw an example in chapter 2 of a variational theorem. It stated that the true eigenvectors of a Hermitian operator are the ones that minimize the *variational energy* (which is given by a simple recipe), subject to the restriction that it be orthogonal to the eigenvectors below it.

Although we will not prove it, the variational energy for this eigenvalue problem is

$$E_{var} = \frac{\displaystyle\sum_{(G\lambda)'(G\lambda)} h_{(G\lambda)}{}^* \Theta^k_{(G\lambda),(G\lambda)'}\, h_{(G\lambda)'}}{\displaystyle\sum_{(G\lambda)} h_{(G\lambda)}{}^* h_{(G\lambda)'}}. \quad (5)$$

A close inspection shows that it is similar in form to $(\mathbf{H}, \Theta\mathbf{H})/(\mathbf{H}, \mathbf{H})$, the variational energy of the original differential equation. In fact, this form holds for all Hermitian eigenvalue problems.

From here, the procedure is simple in principle. Beginning with some guess for $h(\mathbf{G}, \lambda)$, the computer calculates the variational energy and updates its guess so as to lower the variational energy. The guesses are enforced to be orthogonal to any eigenvectors that were found previously. Eventually the algorithm converges on the true $h(\mathbf{G}, \lambda)$ and moves on to the next one.

However, in most cases the dielectric function $\varepsilon(\mathbf{r})$ is not continuous, but rather a patchwork of different constant-ε regions. This causes substantial convergence difficulties, which can be overcome with a suitable interpolation scheme. See Meade et al. (1993a) for more on this detail.

In this manner, all of the eigenvalues $(\omega/c)^2$ can be obtained for a given value of **k**. This information allows us to plot the band structure functions $\omega_n(\mathbf{k})$ as in the text.

[1]See, for example, Golub and Van Loan (1989).

References

Ashcroft, N. W., and N. D. Mermin. 1976. *Solid State Physics.* Saunders College, Philadelphia.

Aspnes, D. E. 1982. "Local-field effects and effective medium theory: A microscopic perspective" *Am. J. Phys.* **50**, 104.

Bloembergen, N. 1965. *Nonlinear Optics.* W. A. Benjamin, New York.

Chan, C. T., K. M. Ho, and C. M. Soukoulis. 1991. "Photonic Band Gaps in Experimentally Realizable Periodic Dielectric Structures." *Europhys. Lett.* **16**, 563.

Fowles, Grant R. 1975. *Introduction to Modern Optics.* Dover, New York.

Golub, G., and C. Van Loan. 1989. *Matrix Computations.* Johns Hopkins University Press, Baltimore.

Griffiths, D. J. 1989. *Introduction to Electrodynamics.* Prentice Hall, Englewood Cliffs, N.J.

Hamermesh, Morton. 1962. *Group Theory and Its Application to Physical Problems.* Dover, New York.

Harrison, W. A. 1979. *Solid State Theory.* Dover, New York.

Harrison, W. A. 1980. *Electronic Structure.* Freeman Press, San Francisco.

Hecht and Zajac. 1974. *Optics.* Addison-Wesley, Reading, Mass.

Ho, K. M., C. T. Chan, and C. M. Soukoulis. 1990. "Existence of photonic gaps in periodic dielectric structures." *Phys. Rev. Lett.* **65**, 3152.

Jackson, J. D. 1962. *Classical Electrodynamics.* John Wiley & Sons, New York.

Kittel, C. 1986. *Solid State Physics.* John Wiley & Sons, New York.

Kleppner, D. 1981. "Inhibited spontaneous emission." *Phys. Rev. Lett.* **47**, 233.

Leung, K. M., and Y. F. Liu. 1990. "Full vector calculations of photonic band structures in face-centered cubic dielectric media." *Phys. Rev. Lett.* **65**, 2646.

Liboff, R. L. 1992. *Introductory Quantum Mechanics.* 2d ed. Addison-Wesley, Reading, Mass.

McCall, S. L., P. M. Platzman, R. Dalichaouch, D. Smith, and S. Schultz. 1991. "Microwave propagation in two-dimensional dielectric lattices." *Phys. Rev. Lett.* **67**, 2017.

Martorell, J., and N. M. Lawandy. 1990. "Observation of inhibited spontaneous emission in a periodic dielectric structure." *Phys. Rev. Lett.* **65**, 1877.

Mathews, J., and R. Walker. 1964. *Mathematical Methods of Physics.* Addison-Wesley, Redwood City, Calif.

Meade, R. D., K. D. Brommer, A. M. Rappe, and J. D. Joannopoulos. 1991a. "Electromagnetic Bloch waves at the surface of a photonic crystal." *Phys. Rev. B.* **44**, 10961.

Meade, R. D., K. D. Brommer, A. M. Rappe, and J. D. Joannopoulos. 1991b. "Photonic bound states in periodic dielectric materials." *Phys. Rev. B* **44**, 13772.

Meade, R. D., K. D. Brommer, A. M. Rappe, and J. D. Joannopoulos. 1992. "Existence of a photonic band gap in two dimensions." *Appl. Phys. Lett.* **61**, 495.

Meade, R. D., K. D. Brommer, A. M. Rappe, J. D. Joannopoulos, and O. L. Alerhand. 1993a. "Accurate theoretical analysis of photonic band gap materials." *Phys. Rev. B* **48**, 8434

Meade, R. D., O. Alerhand, and J. D. Joannopoulos. 1993b. *Handbook of Photonic Band Gap Materials.* JAMteX I.T.R.

Merzbacher, E. 1961. *Quantum Mechanics.* John Wiley & Sons, New York.

Pankove, J. I. 1971. *Optical Processes in Semiconductors.* Dover, New York.

Plihal, M., and A. A. Maradudin. 1991. "Photonic band structure of two-dimensional systems: The triangular lattice." *Phys. Rev. B* **44**, 8565.

Robertson, W. M., G. Arjavalingam, R. D. Meade, K. D. Brommer, A. M. Rappe, and J. D. Joannopoulos. 1992. "Measurement of photonic band structure in a two-dimensional periodic dielectric array." *Phys. Rev. Lett.* **68**, 2023.

Robertson, W. M., G. Arjavalingam, R. D. Meade, K. D. Brommer, A. M. Rappe, and J. D. Joannopoulos. 1993. "Observation of surface photons on periodic dielectric arrays." *Optics Letters* **18**, 528.

Sakurai, J. J. 1985. *Modern Quantum Mechanics.* Addison-Wesley, Reading, Mass.

Satpathy, S., Z. Zhang, and M. Salehpour. 1990. "Theory of photon bands in three-dimensional periodic dielectric structures." *Phys. Rev. Lett.* **64**, 1239.

Shankar, R. 1982. *Principles of Quantum Mechanics.* Plenum Press, New York.

Smith, D. R., R. Dalichaouch, N. Kroll, S. Schultz, S. L. McCall, and P. M. Platzman. 1993. "Photonic band structure and defects in one and two dimensions." *J. Opt. Soc. Am. B* **10**, 314.

Sözüer, H. S., J. W. Haus, and R. Inguva. 1992. "Photonic bands: Convergence problems with the plane-wave method." *Phys. Rev. B* **45**, 13962.

Sze, S. M. 1981. *Physics of Semiconductor Devices.* John Wiley & Sons, New York.

Villeneuve, P., and M. Piche. 1992. "Photonic band gaps in two-dimensional square and hexagonal lattices." *Phys. Rev. B* **46**, 4969.

Wendt, J. R., G. A. Vawter, P. L. Gourley, T. M. Brennan, and B. E. Hammons. 1993. "Nanofabrication of photonic lattice structures in GaAs/AlGaAs." *J. Vac. Sci. & Tech. B* **11**, 2637.

Winn, J. N., R. D. Meade, J. D. Joannopoulos. 1994. "Two-dimensional photonic band gap materials." *J. Mod. Optics.* **41**, 257.

Yablonovitch, E. 1987. "Inhibited spontaneous emission in solid state physics and electronics." *Phys. Rev. Lett.* **58**, 2059.

Yablonovitch, E., and T. J. Gmitter. 1989. "Photonic band structures: The face-centered cubic case." *Phys. Rev. Lett.* **63**, 1950.

Yablonovitch, E., T. J. Gmitter, and K. M. Leung. 1991a. "Photonic band structures: The face-centered cubic case employing non-spherical atoms." *Phys. Rev. Lett.* **67**, 2295.

Yablonovitch, E., T. J. Gmitter, R. D. Meade, K. D. Brommer, A. M. Rappe, and J. D. Joannopoulos. 1991b. "Donor and acceptor modes in photonic band structure." *Phys. Rev. Lett.* **67** 3380.

Yariv, A. 1985. *Optical Electronics.* Holt, Reinhart and Winston, New York.

Yeh, P. 1988. *Optical Waves in Layered Media.* John Wiley & Sons, New York.

Zhang, Ze, and Sashi Satpathy. 1990. "Electromagnetic wave propagation in periodic structures: Bloch wave solutions of Maxwell's equations." *Phys. Rev. Lett.* **65**, 2650.

Index